PLUS: Official SQA Specimen Paper & 2015 Past Paper With Answers

Higher for CfE

Physics

2014 Specimen Question Paper, Model Papers & 2015 Exam

This book contains the official 2014 SQA Specimen Question Paper and 2015 Exam for Higher for CfE Physics, with associated SQA approved answers modified from the official marking instructions that accompany the paper.

In addition the book contains model papers, together with answers, plus study skills advice. These papers, some of which may include a limited number of previously published SQA questions, have been specially commissioned by Hodder Gibson, and have been written by experienced senior teachers and examiners in line with the new Higher for CfE syllabus and assessment outlines, Spring 2014. This is not SQA material but has been devised to provide further practice for Higher for CfE examinations in 2015 and beyond.

Hodder Gibson is grateful to the copyright holders, as credited on the final page of the Answer Section, for permission to use their material. Every effort has been made to trace the copyright holders and to obtain their permission for the use of copyright material. Hodder Gibson will be happy to receive information allowing us to rectify any error or omission in future editions.

Hachette UK's policy is to use papers that are natural, renewable and recyclable products and made from wood grown in sustainable forests. The logging and manufacturing processes are expected to conform to the environmental regulations of the country of origin.

Orders: please contact Bookpoint Ltd, 130 Park Drive, Milton Park, Abingdon, Oxon OX14 4SE. Telephone: (44) 01235 827720. Fax: (44) 01235 400454. Lines are open 9.00–5.00, Monday to Saturday, with a 24-hour message answering service. Visit our website at www.hoddereducation.co.uk. Hodder Gibson can be contacted direct on: Tel: 0141 848 1609; Fax: 0141 889 6315; email: hoddergibson@hodder.co.uk

This collection first published in 2015 by
Hodder Gibson, an imprint of Hodder Education,
An Hachette UK Company
2a Christie Street
Paisley PA1 1NB

Typeset by Aptara, Inc.

Printed in the UK

A catalogue record for this title is available from the British Library

ISBN: 978-1-4718-6081-2

3 2 1

2016 2015

Introduction

Study Skills – what you need to know to pass exams!

Pause for thought

Many students might skip quickly through a page like this. After all, we all know how to revise. Do you really though?

Think about this:

"IF YOU ALWAYS DO WHAT YOU ALWAYS DO, YOU WILL ALWAYS GET WHAT YOU HAVE ALWAYS GOT."

Do you like the grades you get? Do you want to do better? If you get full marks in your assessment, then that's great! Change nothing! This section is just to help you get that little bit better than you already are.

There are two main parts to the advice on offer here. The first part highlights fairly obvious things but which are also very important. The second part makes suggestions about revision that you might not have thought about but which WILL help you.

Part 1

DOH! It's so obvious but ...

Start revising in good time

Don't leave it until the last minute – this will make you panic.

Make a revision timetable that sets out work time AND play time.

Sleep and eat!

Obvious really, and very helpful. Avoid arguments or stressful things too – even games that wind you up. You need to be fit, awake and focused!

Know your place!

Make sure you know exactly **WHEN and WHERE** your exams are.

Know your enemy!

Make sure you know what to expect in the exam.

How is the paper structured?

How much time is there for each question?

What types of question are involved?

Which topics seem to come up time and time again?

Which topics are your strongest and which are your weakest?

Are all topics compulsory or are there choices?

Learn by DOING!

There is no substitute for past papers and practice papers – they are simply essential! Tackling this collection of papers and answers is exactly the right thing to be doing as your exams approach.

Part 2

People learn in different ways. Some like low light, some bright. Some like early morning, some like evening or night. Some prefer warm, some prefer cold. But everyone uses their BRAIN and the brain works when it is active. Passive learning – sitting gazing at notes – is the most INEFFICIENT way to learn anything. Below you will find tips and ideas for making your revision more effective and maybe even more enjoyable. What follows gets your brain active, and active learning works!

Activity 1 – Stop and review

Step 1

When you have done no more than 5 minutes of revision reading STOP!

Step 2

Write a heading in your own words which sums up the topic you have been revising.

Step 3

Write a summary of what you have revised in no more than two sentences. Don't fool yourself by saying, "I know it, but I cannot put it into words". That just means you don't know it well enough. If you cannot write your summary, revise that section again, knowing that you must write a summary at the end of it. Many of you will have notebooks full of blue/black ink writing. Many of the pages will not be especially attractive or memorable so try to liven them up a bit with colour as you are reviewing and rewriting. **This is a great memory aid, and memory is the most important thing.**

Activity 2 – Use technology!

Why should everything be written down? Have you thought about "mental" maps, diagrams, cartoons and colour to help you learn? And rather than write down notes, why not record your revision material?

What about having a text message revision session with friends? Keep in touch with them to find out how and what they are revising and share ideas and questions.

Why not make a video diary where you tell the camera what you are doing, what you think you have learned and what you still have to do? No one has to see or hear it, but the process of having to organise your thoughts in a formal way to explain something is a very important learning practice.

Be sure to make use of electronic files. You could begin to summarise your class notes. Your typing might be slow, but it will get faster and the typed notes will be easier to read than the scribbles in your class notes. Try to add different fonts and colours to make your work stand out. You can easily Google relevant pictures, cartoons and diagrams which you can copy and paste to make your work more attractive and **MEMORABLE**.

Activity 3 – This is it. Do this and you will know lots!

Step 1

In this task you must be very honest with yourself! Find the SQA syllabus for your subject (www.sqa.org.uk). Look at how it is broken down into main topics called MANDATORY knowledge. That means stuff you MUST know.

Step 2

BEFORE you do ANY revision on this topic, write a list of everything that you already know about the subject. It might be quite a long list but you only need to write it once. It shows you all the information that is already in your long-term memory so you know what parts you do not need to revise!

Step 3

Pick a chapter or section from your book or revision notes. Choose a fairly large section or a whole chapter to get the most out of this activity.

With a buddy, use Skype, Facetime, Twitter or any other communication you have, to play the game "If this is the answer, what is the question?". For example, if you are revising Geography and the answer you provide is "meander", your buddy would have to make up a question like "What is the word that describes a feature of a river where it flows slowly and bends often from side to side?".

Make up 10 "answers" based on the content of the chapter or section you are using. Give this to your buddy to solve while you solve theirs.

Step 4

Construct a wordsearch of at least 10 × 10 squares. You can make it as big as you like but keep it realistic. Work together with a group of friends. Many apps allow you to make wordsearch puzzles online. The words and phrases can go in any direction and phrases can be split. Your puzzle must only contain facts linked to the topic you are revising. Your task is to find 10 bits of information to hide in your puzzle, but you must not repeat information that you used in Step 3. DO NOT show where the words are. Fill up empty squares with random letters. Remember to keep a note of where your answers are hidden but do not show your friends. When you have a complete puzzle, exchange it with a friend to solve each other's puzzle.

Step 5

Now make up 10 questions (not "answers" this time) based on the same chapter used in the previous two tasks. Again, you must find NEW information that you have not yet used. Now it's getting hard to find that new information! Again, give your questions to a friend to answer.

Step 6

As you have been doing the puzzles, your brain has been actively searching for new information. Now write a NEW LIST that contains only the new information you have discovered when doing the puzzles. Your new list is the one to look at repeatedly for short bursts over the next few days. Try to remember more and more of it without looking at it. After a few days, you should be able to add words from your second list to your first list as you increase the information in your long-term memory.

FINALLY! Be inspired...

Make a list of different revision ideas and beside each one write **THINGS I HAVE** tried, **THINGS I WILL** try and **THINGS I MIGHT** try. Don't be scared of trying something new.

And remember – "FAIL TO PREPARE AND PREPARE TO FAIL!"

Higher Physics

Assessment

The examination

Section 1

There are 20 multiple choice questions – ensure you answer all of these, even if it means guessing an answer.

Section 2

This section consists of restricted and extended response questions. The mark scored out of **110** here will be scaled to out of **80** and then added to the mark scored in Section 1, giving a total out of **100**.

The majority of the marks will be awarded for applying **knowledge and understanding**. The other marks will be awarded for **applying scientific inquiry, scientific analytical thinking and problem solving skills**.

The assignment

The evidence will be submitted to SQA for external marking. This will be marked out of **20**.

The final mark for Higher Physics will out of a total of 120.

The examination – general points

Standard 3-marker

Look out for these. The breakdown of the marks will be:

1 mark – selecting equation

1 mark – substitution

1 mark – answer, including unit.

Do not rearrange equations in algebraic form. Select the appropriate equation, substitute the given values, and then rearrange the equation to obtain the required unknown. **This minimises the risk of wrong substitution.**

For example:

Calculate the acceleration of a mass of 5 kg when acted on by a resultant force of 10 N.

Solution 1	**Solution 2**	**Solution 3**
F = ma (1)	F = ma (1)	F = ma (1)
10 = 5 a (1)	a = m/F = 5/10	10 = 5 a (1)
a = 2 ms^{-2} (1)	= 0.5 ms^{-2}	= 0.5 ms^{-2}
3 marks	**1 mark for selecting formula.**	**2 marks (1 mark for selecting formula 1 mark for correct substitution.)**

Use of the data sheet and data booklet

Clearly show where you have substituted a value from the data sheet. Do not leave *Ho* in an equation. You must show the value of *Ho* has been correctly substituted.

Rounding – do not round the given values.

E.g. mass of a proton = 1.673×10^{-27} kg

NOT 1.67×10^{-27} kg.

Although many of the required equations are given, **it is better to know the basic equations to gain time in the examination**.

"Show" questions

Generally **all steps** for these must be given. **Do not assume that substitutions are obvious to the marker.** All equations used must be stated separately and then clearly substituted if required. Many candidates will look at the end product and somehow end up with the required answer. The marker has to ensure that the path to the solution is clear. It is good practice to state why certain equations are used, explaining the Physics behind them.

Make sure you include the unit in the final answer.

Definitions

Know and understand definitions given in the course. Definitions often come from the interpretation of an equation.

For example, 1 Farad is equivalent to a $1CV^{-1}$ (Q = CV)

Diagrams, graphs and sketch graphs

When drawing diagrams, use a ruler and use appropriate labels. Angles will be important in certain diagrams. Too many candidates attempt to draw ray diagrams freehand.

When drawing graphs, use a ruler and pencil to draw for axes. Label the axes correctly including units and the origin.

When tackling sketch graphs, care should be taken to be as neat as possible. Ensure axes are drawn in pencil with a ruler. Also ensure you use a ruler to draw a straight line graph.

Significant figures

Do not round off in intermediate calculations, but round off in the final answer to an appropriate number of figures.

Rounding off to three significant figures in the final answer will generally be acceptable.

Prefixes

Ensure you know all the prefixes required and be able to convert them to the correct power of 10.

"Explain"/descriptive questions

These tend to be done poorly. Ensure all points are covered and read over again in order to check there are no mistakes. Try to be clear and to the point, highlighting the relevant Physics.

Do not use up and down arrows in a description – this may help you in shorthand, but these must be translated to words.

Be aware some answers require justification. No attempt at a justification can mean no marks awarded.

Two or more attempts at an answer

The attempt that the candidate does not want to be considered should be scored out. Otherwise zero marks could be awarded.

Do not be attempted to give extra information that might be incorrect – marks could be deducted for each incorrect piece of information. This might include converting incorrectly m to nm in the last line of an answer, when it is not required.

At the end of the exam, if you have time, quickly go over each answer and make sure you have the correct unit inserted.

Skills

Experimental descriptions/planning

You could well be called on to describe an experimental set up.

Ensure your description is clear enough for another person to repeat it.

Include a clearly labelled diagram.

Suggested improvements to experimental procedure

Look at the percentage uncertainties in the measured quantities and decide which is most significant. Suggest how the size of this uncertainty could be reduced – do not suggest use better apparatus! It might be better to repeat readings, so random uncertainty is reduced or increase distances to reduce the percentage uncertainty in scale reading. There could be a systematic uncertainty that is affecting all readings. It really depends on the experiment. Use your judgement.

Handling data

Relationships

There are two methods to prove a direct or inverse relationship.

Graphical approach

Plot the graphs with the appropriate values x and y values and look for a straight line – better plotted in pencil in case of mistakes. **Do not force a line through the origin!** (A vs B for a direct relationship, C vs I / D for an inverse relationship)

Algebraic approach

If it appears that $A \propto B$ then calculate the value of A / B **for all values**.

If these show that A/B = k then the relationship holds.

If it appears that $C \propto 1/D$ then calculate the value C.D **for all values**.

If these show that C.D = k then the relationship holds.

Using the equation of a straight line, $y = mx + c$.

Be aware that the gradient of the line can often lead to required values.

For example, finding the internal resistance and emf of a cell.

$E = IR + Ir = V + Ir$

$V = -rI + E$ in the form of $y = mx + c$

By plotting the graph of V against I, the value of the gradient will give $-r$ and the intercept will give E.

Ensure you are clear on how to calculate the gradient of a line.

Unfamiliar content

If you come across unfamiliar content such as an equation or measurements from an unfamiliar experiment – don't panic! Just read the instructions. Relationships between the quantities can be found graphically or algebraically.

Uncertainties

In this section you need to understand the following:

Systematic, scale reading (analogue and digital) and random uncertainties.

Percentage, absolute uncertainties.

Percentage uncertainty in final answer is taken as the largest percentage uncertainty in the components.

E.g. $V = (7.5 \pm 0.1)$ V	$I = (0.85 \pm 0.05)$ A
$= 7.5$V $\pm 1.3\%$	$= 0.85$ A $\pm 5.8\%$
$R = V/I = 7.5/0.85 = 8.8\,\Omega + 5.8\% = (8.8 \pm 0.1)\,\Omega$	

Open-ended questions

There will generally be two open ended questions in the paper worth 3 marks each. Some candidates look upon these as mini essays. Remember that they are only worth 3 marks and it gives the opportunity to demonstrate knowledge and understanding. However, do not spend too long on these. It might be better to revisit them at the end of the exam.

Some students prefer to use bullet points to highlight the main areas of understanding.

Ensure you reread the question and understand exactly what is being asked.

Once you have written your response, read over it again to ensure it makes sense.

Good luck!

Remember that the rewards for passing Higher Physics are well worth it! Your pass will help you get the future you want for yourself. In the exam, be confident in your own ability. If you're not sure how to answer a question, trust your instincts and just give it a go anyway – keep calm and don't panic! GOOD LUCK!

HIGHER FOR CfE

2014 Specimen Question Paper

National
Qualifications
SPECIMEN ONLY

SQ37/H/02

Physics
Section 1—Questions

Date — Not applicable

Duration — 2 hours and 30 minutes

Instructions for the completion of Section 1 are given on *Page two* of your question and answer booklet SQ37/H/01.

Record your answers on the answer grid on *Page three* of your question and answer booklet.

Reference may be made to the Data Sheet on *Page two* of this booklet and to the Relationships Sheet SQ37/H/11.

Before leaving the examination room you must give your question and answer booklet to the Invigilator; if you do not, you may lose all the marks for this paper.

DATA SHEET

COMMON PHYSICAL QUANTITIES

Quantity	*Symbol*	*Value*	*Quantity*	*Symbol*	*Value*
Speed of light in vacuum	c	$3{\cdot}00 \times 10^{8}$ m s^{-1}	Planck's constant	h	$6{\cdot}63 \times 10^{-34}$ J s
Magnitude of the charge on an electron	e	$1{\cdot}60 \times 10^{-19}$ C	Mass of electron	m_e	$9{\cdot}11 \times 10^{-31}$ kg
Universal Constant of Gravitation	G	$6{\cdot}67 \times 10^{-11}$ m^{3} kg^{-1} s^{-2}	Mass of neutron	m_n	$1{\cdot}675 \times 10^{-27}$ kg
Gravitational acceleration on Earth	g	$9{\cdot}8$ m s^{-2}	Mass of proton	m_p	$1{\cdot}673 \times 10^{-27}$ kg
Hubble's constant	H_0	$2{\cdot}3 \times 10^{-18}$ s^{-1}			

REFRACTIVE INDICES

The refractive indices refer to sodium light of wavelength 589 nm and to substances at a temperature of 273 K.

Substance	*Refractive index*	*Substance*	*Refractive index*
Diamond	2·42	Water	1·33
Crown glass	1·50	Air	1·00

SPECTRAL LINES

Element	*Wavelength*/nm	*Colour*
Hydrogen	656	Red
	486	Blue-green
	434	Blue-violet
	410	Violet
	397	Ultraviolet
	389	Ultraviolet
Sodium	589	Yellow

Element	*Wavelength*/nm	*Colour*
Cadmium	644	Red
	509	Green
	480	Blue

Lasers

Element	*Wavelength*/nm	*Colour*
Carbon dioxide	9550 } 10590 }	Infrared
Helium-neon	633	Red

PROPERTIES OF SELECTED MATERIALS

Substance	*Density*/kg m^{-3}	*Melting Point*/K	*Boiling Point*/K
Aluminium	$2{\cdot}70 \times 10^{3}$	933	2623
Copper	$8{\cdot}96 \times 10^{3}$	1357	2853
Ice	$9{\cdot}20 \times 10^{2}$	273	....
Sea Water	$1{\cdot}02 \times 10^{3}$	264	377
Water	$1{\cdot}00 \times 10^{3}$	273	373
Air	1·29	....	....
Hydrogen	$9{\cdot}0 \times 10^{-2}$	14	20

The gas densities refer to a temperature of 273 K and a pressure of $1{\cdot}01 \times 10^{5}$ Pa.

SECTION 1 — 20 marks

Attempt ALL questions

1. A trolley has a constant acceleration of $3\,m\,s^{-2}$. This means that

A the distance travelled by the trolley increases by 3 metres per second every second

B the displacement of the trolley increases by 3 metres per second every second

C the speed of the trolley is $3\,m\,s^{-1}$ every second

D the velocity of the trolley is $3\,m\,s^{-1}$ every second

E the velocity of the trolley increases by $3\,m\,s^{-1}$ every second.

2. Which of the following velocity-time graphs represents the motion of an object that changes direction?

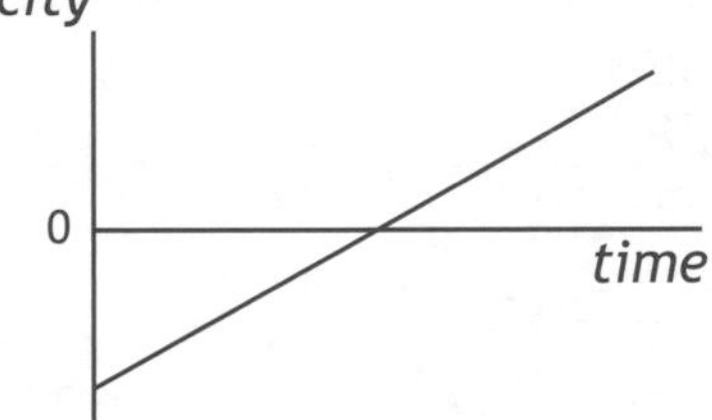

B

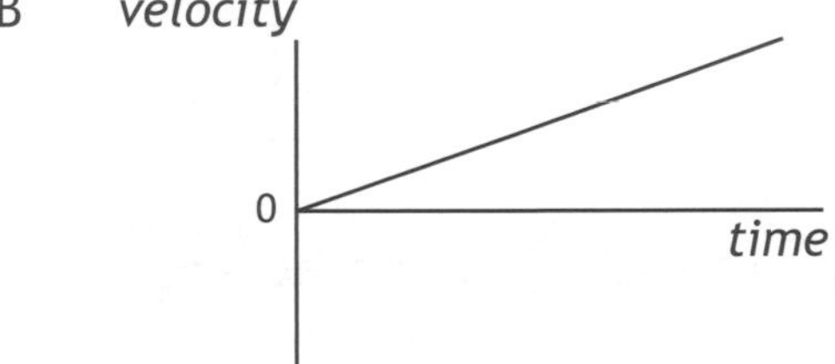

C

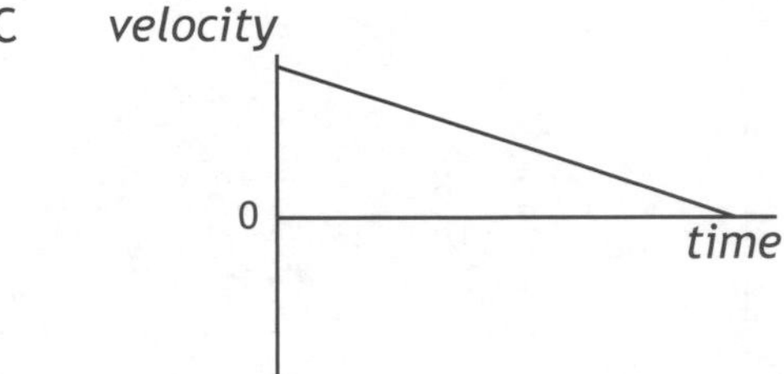

D

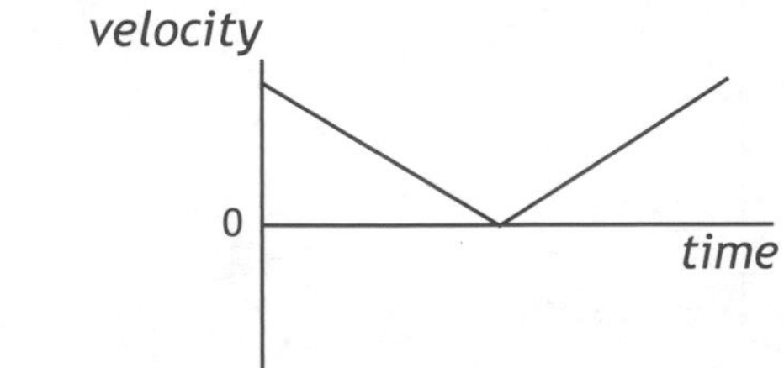

E

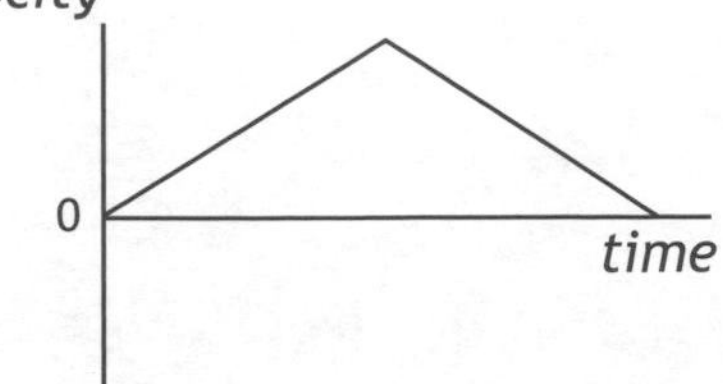

3. A football of mass 0·75 kg is initially at rest. A girl kicks the football and it moves off with an initial speed of $12\,\text{m s}^{-1}$. The time of contact between the girl's foot and the football is 0·15 s.

The average force applied to the football as it is kicked is

A 1·4 N

B 1·8 N

C 2·4 N

D 60 N

E 80 N.

4. Two small asteroids are 12 m apart.

The masses of the asteroids are $2{\cdot}0 \times 10^3$ kg and $0{\cdot}050 \times 10^3$ kg.

The gravitational force acting between the asteroids is

A $1{\cdot}2 \times 10^{-9}$ N

B $4{\cdot}6 \times 10^{-8}$ N

C $5{\cdot}6 \times 10^{-7}$ N

D $1{\cdot}9 \times 10^{-6}$ N

E $6{\cdot}8 \times 10^{3}$ N.

5. A spaceship on a launch pad is measured to have a length L. This spaceship has a speed of $2{\cdot}5 \times 10^8\,\text{m s}^{-1}$ as it passes a planet.

Which row in the table describes the length of the spaceship as measured by the pilot in the spaceship and an observer on the planet?

	Length measured by pilot in the spaceship	*Length measured by observer on the planet*
A	L	less than L
B	L	greater than L
C	L	L
D	less than L	L
E	greater than L	L

6. The siren on an ambulance is emitting sound with a constant frequency of 900 Hz. The ambulance is travelling at a constant speed of $25\,m\,s^{-1}$ as it approaches and passes a stationary observer. The speed of sound in air is $340\,m\,s^{-1}$.

Which row in the table shows the frequency of the sound heard by the observer as the ambulance approaches and as it moves away from the observer?

	Frequency as ambulance approaches (Hz)	*Frequency as ambulance moves away* (Hz)
A	900	900
B	971	838
C	838	900
D	971	900
E	838	971

7. The photoelectric effect

A is evidence for the wave nature of light

B can be observed using a diffraction grating

C can only be observed with ultra-violet light

D can only be observed with infra-red light

E is evidence for the particulate nature of light.

8. A ray of red light is incident on a glass block as shown.

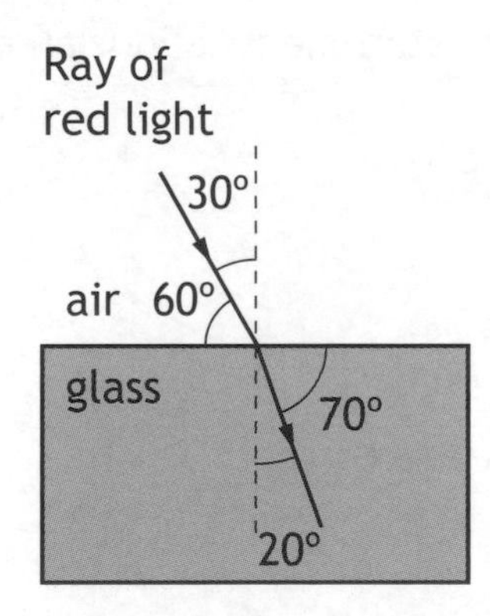

The refractive index of the glass for this light is

A 0·53

B 0·68

C 1·46

D 1·50

E 2·53.

9. A ray of red light travels from air into water.

Which row in the table describes the change, if any, in speed and frequency of a ray of red light as it travels from air into water?

	Speed	*Frequency*
A	increases	increases
B	increases	stays constant
C	decreases	stays constant
D	decreases	decreases
E	stays constant	decreases

10. Light from a point source is incident on a screen. The screen is 3·0 m from the source.

The irradiance at the screen is $8{\cdot}0\,W\,m^{-2}$.

The light source is now moved to a distance of 12 m from the screen.

The irradiance at the screen is now

A $0{\cdot}50\,W\,m^{-2}$

B $1{\cdot}0\,W\,m^{-2}$

C $2{\cdot}0\,W\,m^{-2}$

D $4{\cdot}0\,W\,m^{-2}$

E $8{\cdot}0\,W\,m^{-2}$.

11. A student makes the following statements about an electron.

I An electron is a boson.

II An electron is a lepton.

III An electron is a fermion.

Which of these statements is/are correct?

A I only

B II only

C III only

D I and II only

E II and III only

12. Radiation of frequency $9{\cdot}40 \times 10^{14}$ Hz is incident on a clean metal surface.

The work function of the metal is $3{\cdot}78 \times 10^{-19}$ J.

The maximum kinetic energy of an emitted photoelectron is

A $2{\cdot}45 \times 10^{-19}$ J

B $3{\cdot}78 \times 10^{-19}$ J

C $6{\cdot}23 \times 10^{-19}$ J

D $1{\cdot}00 \times 10^{-18}$ J

E $2{\cdot}49 \times 10^{33}$ J.

13. The diagram represents the electric field around a single point charge.

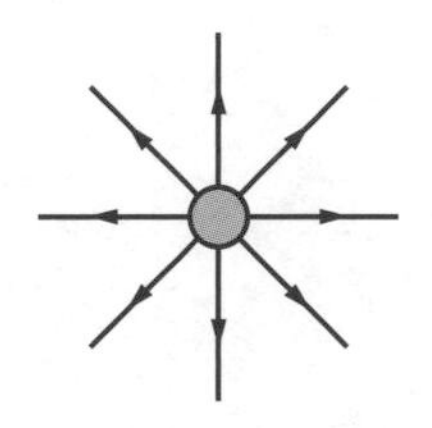

A student makes the following statements about this diagram.

I The separation of the field lines indicates the strength of the field.

II The arrows on the field lines indicate the direction in which an electron would move if placed in the field.

III The point charge is positive.

Which of these statements is/are correct?

A I only

B II only

C I and III only

D II and III only

E I, II and III

14. In the diagrams below, each resistor has the same resistance.

Which combination has the least value of the effective resistance between the terminals X and Y?

A
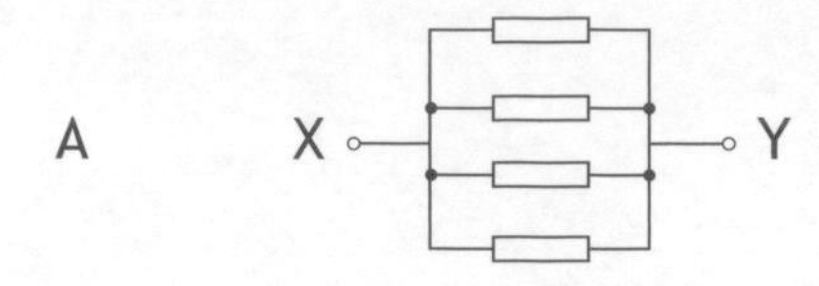

B
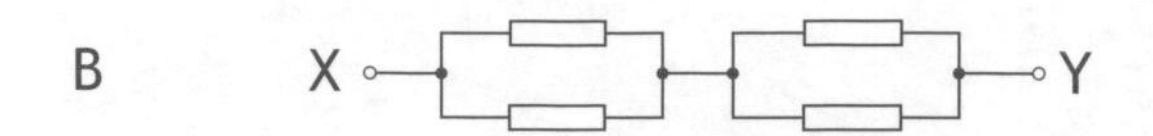

C
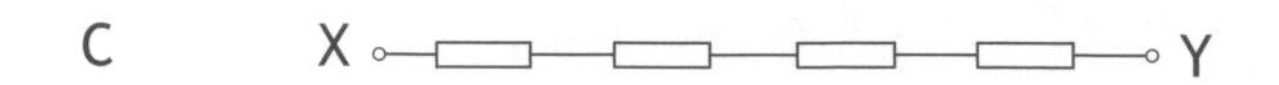

D
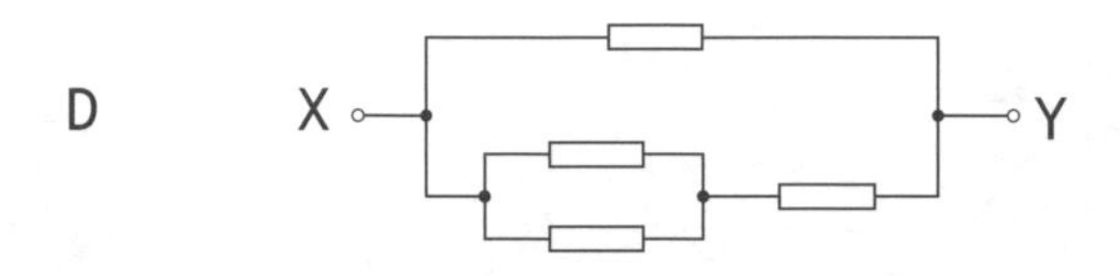

E X Y

15. A student makes the following statements about charges in electric fields.

I An electric field applied to a conductor causes the free electric charges in the conductor to move.

II When a charge is moved in an electric field work is done.

III An electric charge experiences a force in an electric field.

Which of these statements is/are correct?

A II only

B III only

C I and II only

D II and III only

E I, II and III

16. A circuit is set up as shown.

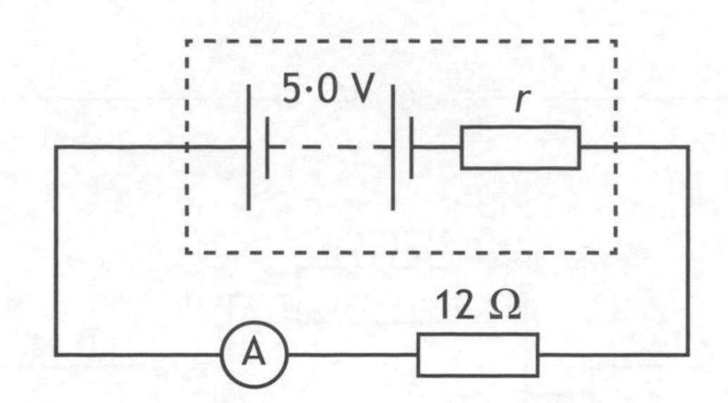

The e.m.f. of the battery is 5·0 V.

The reading on the ammeter is 0·35 A.

The internal resistance *r* of the battery is

A 0·28 Ω

B 0·80 Ω

C 1·15 Ω

D 2·3 Ω

E 3·2 Ω.

17. The e.m.f. of a battery is

A the total energy supplied by the battery

B the voltage lost due to the internal resistance of the battery

C the total charge that passes through the battery

D the number of coulombs of charge passing through the battery per second

E the energy supplied to each coulomb of charge passing through the battery.

18. The r.m.s. voltage of the mains supply is 230 V.

The approximate value of the peak voltage is

A 115 V

B 163 V

C 325 V

D 460 V

E 651 V.

19. Four resistors each of resistance 20 Ω are connected to a 60 V supply of negligible internal resistance as shown.

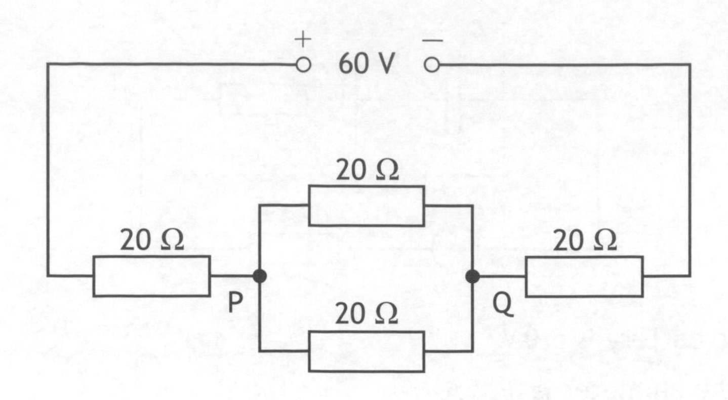

The potential difference across PQ is

A 12 V

B 15 V

C 20 V

D 24 V

E 30 V.

20. Photons with a frequency of $4{\cdot}57 \times 10^{14}$ Hz are incident on a p-n junction in a solar cell.

The maximum potential difference these photons produce across this junction is

A 1·34 V

B 1·89 V

C 2·67 V

D 3·79 V

E 5·34 V.

[END OF SECTION 1. NOW ATTEMPT THE QUESTIONS IN SECTION 2 OF YOUR QUESTION AND ANSWER BOOKLET]

SQ37/H/11

Physics
Relationships Sheet

Date — Not applicable

Relationships required for Physics Higher

$d = \bar{v}t$

$s = \bar{v}t$

$v = u + at$

$s = ut + \frac{1}{2}at^2$

$v^2 = u^2 + 2as$

$s = \frac{1}{2}(u + v)t$

$W = mg$

$F = ma$

$E_W = Fd$

$E_p = mgh$

$E_k = \frac{1}{2}mv^2$

$P = \frac{E}{t}$

$p = mv$

$Ft = mv - mu$

$F = G\frac{m_1 m_2}{r^2}$

$t' = \frac{t}{\sqrt{1 - \left(\frac{v}{c}\right)^2}}$

$l' = l\sqrt{1 - \left(\frac{v}{c}\right)^2}$

$f_o = f_s\left(\frac{v}{v \pm v_s}\right)$

$z = \frac{\lambda_{observed} - \lambda_{rest}}{\lambda_{rest}}$

$z = \frac{v}{c}$

$v = H_0 d$

$W = QV$

$E = mc^2$

$E = hf$

$E_k = hf - hf_0$

$E_2 - E_1 = hf$

$T = \frac{1}{f}$

$v = f\lambda$

$d\sin\theta = m\lambda$

$n = \frac{\sin\theta_1}{\sin\theta_2}$

$\frac{\sin\theta_1}{\sin\theta_2} = \frac{\lambda_1}{\lambda_2} = \frac{v_1}{v_2}$

$\sin\theta_c = \frac{1}{n}$

$I = \frac{k}{d^2}$

$I = \frac{P}{A}$

$V_{peak} = \sqrt{2}V_{rms}$

$I_{peak} = \sqrt{2}I_{rms}$

$Q = It$

$V = IR$

$P = IV = I^2R = \frac{V^2}{R}$

$R_T = R_1 + R_2 + \ldots.$

$\frac{1}{R_T} = \frac{1}{R_1} + \frac{1}{R_2} + \ldots.$

$E = V + Ir$

$V_1 = \left(\frac{R_1}{R_1 + R_2}\right)V_s$

$\frac{V_1}{V_2} = \frac{R_1}{R_2}$

$C = \frac{Q}{V}$

$E = \frac{1}{2}QV = \frac{1}{2}CV^2 = \frac{1}{2}\frac{Q^2}{C}$

path difference $= m\lambda$ or $\left(m + \frac{1}{2}\right)\lambda$ where $m = 0, 1, 2 \ldots$

random uncertainty $= \frac{\text{max. value} - \text{min. value}}{\text{number of values}}$

Additional Relationships

Circle

circumference = $2\pi r$

area = πr^2

Sphere

area = $4\pi r^2$

volume = $\frac{4}{3}\pi r^3$

Trigonometry

$\sin\Theta = \frac{\text{opposite}}{\text{hypotenuse}}$

$\cos\Theta = \frac{\text{adjacent}}{\text{hypotenuse}}$

$\tan\Theta = \frac{\text{opposite}}{\text{adjacent}}$

$\sin^2\Theta + \cos^2\Theta = 1$

Electron Arrangements of Elements

Key

Atomic number
Symbol
Electron arrangement
Name

Transition Elements

Group 1 (1)	Group 2 (2)	(3)	(4)	(5)	(6)	(7)	(8)	(9)	(10)	(11)	(12)	Group 3 (13)	Group 4 (14)	Group 5 (15)	Group 6 (16)	Group 7 (17)	Group 0 (18)
1 **H** 1 Hydrogen																	2 **He** 2 Helium
3 **Li** 2,1 Lithium	4 **Be** 2,2 Beryllium											5 **B** 2,3 Boron	6 **C** 2,4 Carbon	7 **N** 2,5 Nitrogen	8 **O** 2,6 Oxygen	9 **F** 2,7 Fluorine	10 **Ne** 2,8 Neon
11 **Na** 2,8,1 Sodium	12 **Mg** 2,8,2 Magnesium											13 **Al** 2,8,3 Aluminium	14 **Si** 2,8,4 Silicon	15 **P** 2,8,5 Phosphorus	16 **S** 2,8,6 Sulfur	17 **Cl** 2,8,7 Chlorine	18 **Ar** 2,8,8 Argon
19 **K** 2,8,8,1 Potassium	20 **Ca** 2,8,8,2 Calcium	21 **Sc** 2,8,9,2 Scandium	22 **Ti** 2,8,10,2 Titanium	23 **V** 2,8,11,2 Vanadium	24 **Cr** 2,8,13,1 Chromium	25 **Mn** 2,8,13,2 Manganese	26 **Fe** 2,8,14,2 Iron	27 **Co** 2,8,15,2 Cobalt	28 **Ni** 2,8,16,2 Nickel	29 **Cu** 2,8,18,1 Copper	30 **Zn** 2,8,18,2 Zinc	31 **Ga** 2,8,18,3 Gallium	32 **Ge** 2,8,18,4 Germanium	33 **As** 2,8,18,5 Arsenic	34 **Se** 2,8,18,6 Selenium	35 **Br** 2,8,18,7 Bromine	36 **Kr** 2,8,18,8 Krypton
37 **Rb** 2,8,18,8,1 Rubidium	38 **Sr** 2,8,18,8,2 Strontium	39 **Y** 2,8,18,9,2 Yttrium	40 **Zr** 2,8,18, 10,2 Zirconium	41 **Nb** 2,8,18, 12,1 Niobium	42 **Mo** 2,8,18,13, 1 Molybdenum	43 **Tc** 2,8,18,13, 2 Technetium	44 **Ru** 2,8,18,15, 1 Ruthenium	45 **Rh** 2,8,18,16, 1 Rhodium	46 **Pd** 2,8,18, 18,0 Palladium	47 **Ag** 2,8,18, 18,1 Silver	48 **Cd** 2,8,18, 18,2 Cadmium	49 **In** 2,8,18, 18,3 Indium	50 **Sn** 2,8,18, 18,4 Tin	51 **Sb** 2,8,18, 18,5 Antimony	52 **Te** 2,8,18, 18,6 Tellurium	53 **I** 2,8,18, 18,7 Iodine	54 **Xe** 2,8,18, 18,8 Xenon
55 **Cs** 2,8,18,18, 8,1 Caesium	56 **Ba** 2,8,18,18, 8,2 Barium	57 **La** 2,8,18,18, 9,2 Lanthanum	72 **Hf** 2,8,18,32, 10,2 Hafnium	73 **Ta** 2,8,18, 32,11,2 Tantalum	74 **W** 2,8,18,32, 12,2 Tungsten	75 **Re** 2,8,18,32, 13,2 Rhenium	76 **Os** 2,8,18,32, 14,2 Osmium	77 **Ir** 2,8,18,32, 15,2 Iridium	78 **Pt** 2,8,18,32, 17,1 Platinum	79 **Au** 2,8,18, 32,18,1 Gold	80 **Hg** 2,8,18, 32,18,2 Mercury	81 **Tl** 2,8,18, 32,18,3 Thallium	82 **Pb** 2,8,18, 32,18,4 Lead	83 **Bi** 2,8,18, 32,18,5 Bismuth	84 **Po** 2,8,18, 32,18,6 Polonium	85 **At** 2,8,18, 32,18,7 Astatine	86 **Rn** 2,8,18, 32,18,8 Radon
87 **Fr** 2,8,18,32, 18,8,1 Francium	88 **Ra** 2,8,18,32, 18,8,2 Radium	89 **Ac** 2,8,18,32, 18,9,2 Actinium	104 **Rf** 2,8,18,32, 32,10,2 Rutherfordium	105 **Db** 2,8,18,32, 32,11,2 Dubnium	106 **Sg** 2,8,18,32, 32,12,2 Seaborgium	107 **Bh** 2,8,18,32, 32,13,2 Bohrium	108 **Hs** 2,8,18,32, 32,14,2 Hassium	109 **Mt** 2,8,18,32, 32,15,2 Meitnerium	110 **Ds** 2,8,18,32, 32,17,1 Darmstadtium	111 **Rg** 2,8,18,32, 32,18,1 Roentgenium	112 **Cn** 2,8,18,32, 32,18,2 Copernicium						

Lanthanides	57 **La** 2,8,18, 18,9,2 Lanthanum	58 **Ce** 2,8,18, 20,8,2 Cerium	59 **Pr** 2,8,18,21, 8,2 Praseodymium	60 **Nd** 2,8,18,22, 8,2 Neodymium	61 **Pm** 2,8,18,23, 8,2 Promethium	62 **Sm** 2,8,18,24, 8,2 Samarium	63 **Eu** 2,8,18,25, 8,2 Europium	64 **Gd** 2,8,18,25, 9,2 Gadolinium	65 **Tb** 2,8,18,27, 8,2 Terbium	66 **Dy** 2,8,18,28, 8,2 Dysprosium	67 **Ho** 2,8,18,29, 8,2 Holmium	68 **Er** 2,8,18,30, 8,2 Erbium	69 **Tm** 2,8,18,31, 8,2 Thulium	70 **Yb** 2,8,18,32, 8,2 Ytterbium	71 **Lu** 2,8,18,32, 9,2 Lutetium
Actinides	89 **Ac** 2,8,18,32, 18,9,2 Actinium	90 **Th** 2,8,18,32, 18,10,2 Thorium	91 **Pa** 2,8,18,32, 20,9,2 Protactinium	92 **U** 2,8,18,32, 21,9,2 Uranium	93 **Np** 2,8,18,32, 22,9,2 Neptunium	94 **Pu** 2,8,18,32, 24,8,2 Plutonium	95 **Am** 2,8,18,32, 25,8,2 Americium	96 **Cm** 2,8,18,32, 25,9,2 Curium	97 **Bk** 2,8,18,32, 27,8,2 Berkelium	98 **Cf** 2,8,18,32, 28,8,2 Californium	99 **Es** 2,8,18,32, 29,8,2 Einsteinium	100 **Fm** 2,8,18,32, 30,8,2 Fermium	101 **Md** 2,8,18,32, 31,8,2 Mendelevium	102 **No** 2,8,18,32, 32,8,2 Nobelium	103 **Lr** 2,8,18,32, 32,9,2 Lawrencium

FOR OFFICIAL USE

H

National
Qualifications
SPECIMEN ONLY

Mark

SQ37/H/01

Physics Section 1— Answer Grid and Section 2

Date — Not applicable

Duration — 2 hours 30 minutes

Fill in these boxes and read what is printed below.

Full name of centre

Town

Forename(s)

Surname

Number of seat

Date of birth

Day Month Year

D D M M Y Y

Scottish candidate number

Total marks — 130

SECTION 1 — 20 marks

Attempt ALL questions.

Instructions for the completion of Section 1 are given on *Page two*.

SECTION 2 — 110 marks

Attempt ALL questions.

Reference may be made to the Data Sheet on *Page two* of the question paper SQ37/H/02 and to the Relationship Sheet SQ37/H/11.

Write your answers clearly in the spaces provided in this booklet. Additional space for answers and rough work is provided at the end of this booklet. If you use this space you must clearly identify the question number you are attempting. Any rough work must be written in this booklet. You should score through your rough work when you have written your final copy.

Use **blue** or **black** ink.

Care should be taken to give an appropriate number of significant figures in the final answers to calculations.

Before leaving the examination room you must give this booklet to the Invigilator; if you do not, you may lose all the marks for this paper.

SECTION 1 — 20 marks

The questions for Section 1 are contained in the question paper SQ37/H/02.
Read these and record your answers on the answer grid on *Page three* opposite.
Do **NOT** use gel pens.

1. The answer to each question is **either** A, B, C, D or E. Decide what your answer is, then fill in the appropriate bubble (see sample question below).

2. There is **only one correct** answer to each question.

3. Any rough working should be done on the additional space for answers and rough work at the end of this booklet.

Sample Question

The energy unit measured by the electricity meter in your home is the:

A ampere
B kilowatt-hour
C watt
D coulomb
E volt.

The correct answer is **B**—kilowatt-hour. The answer **B** bubble has been clearly filled in (see below).

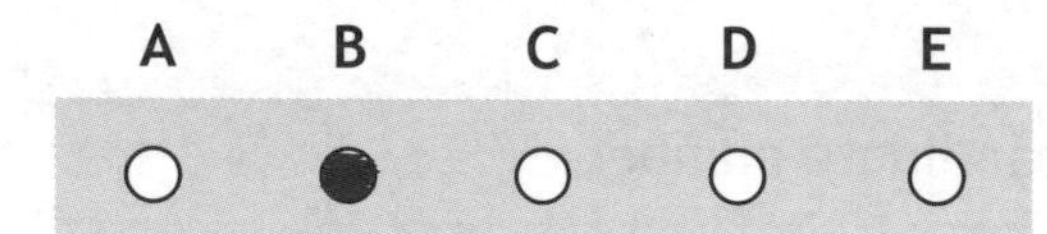

Changing an answer

If you decide to change your answer, cancel your first answer by putting a cross through it (see below) and fill in the answer you want. The answer below has been changed to **D**.

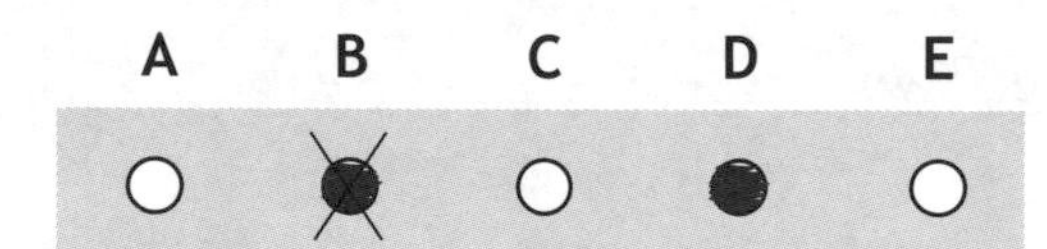

If you then decide to change back to an answer you have already scored out, put a tick (✓) to the **right** of the answer you want, as shown below:

A B C D E

or

A B C D E

SECTION 1 — Answer Grid

	A	B	C	D	E
1	○	○	○	○	○
2	○	○	○	○	○
3	○	○	○	○	○
4	○	○	○	○	○
5	○	○	○	○	○
6	○	○	○	○	○
7	○	○	○	○	○
8	○	○	○	○	○
9	○	○	○	○	○
10	○	○	○	○	○
11	○	○	○	○	○
12	○	○	○	○	○
13	○	○	○	○	○
14	○	○	○	○	○
15	○	○	○	○	○
16	○	○	○	○	○
17	○	○	○	○	○
18	○	○	○	○	○
19	○	○	○	○	○
20	○	○	○	○	○

SECTION 2 — 110 marks

Attempt ALL questions

MARKS | DO NOT WRITE IN THIS MARGIN

1. A golf ball is hit with a velocity of $50{\cdot}0\,m\,s^{-1}$ at an angle of 35° to the horizontal as shown.

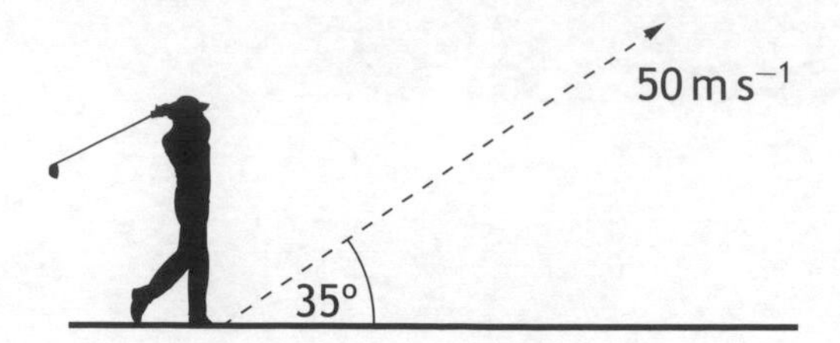

(a) (i) Calculate the horizontal component of the initial velocity of the ball. **1**

Space for working and answer

(ii) Calculate the vertical component of the initial velocity of the ball. **1**

Space for working and answer

MARKS DO NOT WRITE IN THIS MARGIN

1. **(continued)**

(b) The diagram below shows the trajectory of the ball when air resistance is negligible.

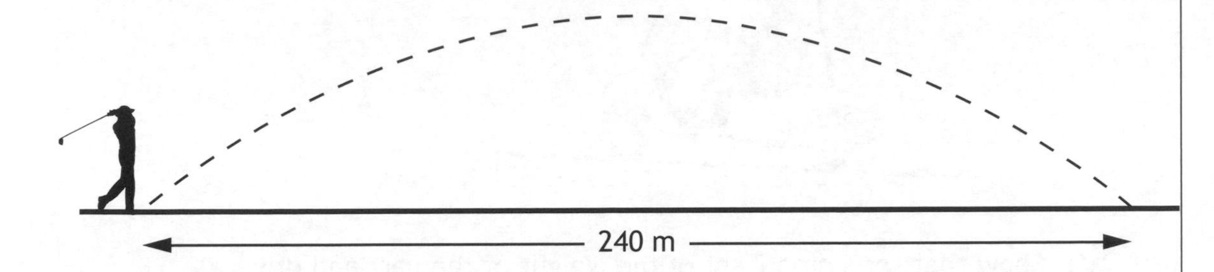

Show that the horizontal distance travelled by the ball is 240 m. 4

Space for working and answer

MARKS | DO NOT WRITE IN THIS MARGIN

2. An electric cart and driver accelerate up a slope. The slope is at an angle of 3·2° to the horizontal. The combined mass of the cart and driver is 220 kg.

(a) (i) Show that the component of the weight of the cart and driver acting down the slope is 120 N. **2**

Space for working and answer

(ii) At one point on the slope the driving force produced by the cart's motor is 230 N and at this point the total frictional force acting on the cart and driver is 48 N.

Calculate the acceleration of the cart and the driver at this point. **4**

Space for working and answer

MARKS DO NOT WRITE IN THIS MARGIN

2. (a) (continued)

(iii) Explain, in terms of the forces, why there is a maximum angle of slope that the cart can ascend. 2

(b) The electric motor in the cart is connected to a battery of e.m.f. 48 V and internal resistance 0·52 Ω.

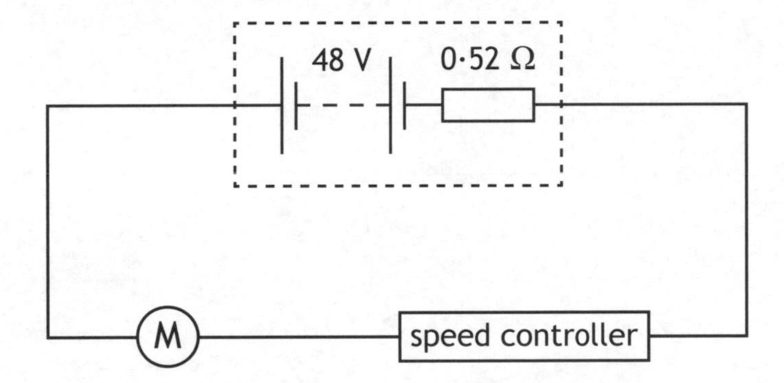

The current in the circuit is 22 A.

(i) Show that the lost volts in the battery is 11 V. 2

Space for working and answer

MARKS | DO NOT WRITE IN THIS MARGIN

2. (b) **(continued)**

(ii) Calculate the output power supplied to the circuit when the current is 22 A. **4**

Space for working and answer

(c) The driving force produced by the cart's motor is now increased.

State what happens to the potential difference across the battery.

You must justify your answer. **3**

MARKS | DO NOT WRITE IN THIS MARGIN

3. When a car brakes, kinetic energy is turned into heat and sound.

In order to make cars more efficient some manufacturers are developing kinetic energy recovery systems (KERS). These systems store some of the energy that would otherwise be lost as heat and sound.

Estimate the maximum energy that could be stored in such a system when a car brakes.

Clearly show your working for the calculation and any estimates you have made. 4

Space for working and answer

MARKS DO NOT WRITE IN THIS MARGIN

4. Muons are sub-atomic particles produced when cosmic rays enter the atmosphere about 10 km above the surface of the Earth.

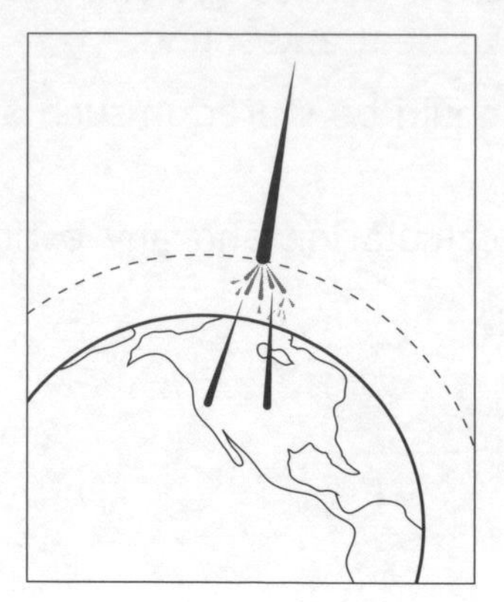

Muons have a mean lifetime of $2{\cdot}2 \times 10^{-6}$ s in their frame of reference. Muons are travelling at $0{\cdot}995c$ relative to an observer on Earth.

(a) Show that the mean distance travelled by the muons in their frame of reference is 660 m. **2**

Space for working and answer

(b) Calculate the mean lifetime of the muons as measured by the observer on Earth. **3**

Space for working and answer

MARKS DO NOT WRITE IN THIS MARGIN

4. **(continued)**

(c) Explain why a greater number of muons are detected on the surface of the Earth than would be expected if relativistic effects were not taken into account. **1**

MARKS DO NOT WRITE IN THIS MARGIN

5. A picture of a helmet designed to be worn when riding a bicycle is shown.

The bicycle helmet has a hard outer shell and a soft expanded polystyrene foam liner.

Using your knowledge of physics, comment on the suitability of this design for a bicycle helmet. 3

MARKS DO NOT WRITE IN THIS MARGIN

6. (a) The diagram below represents part of the emission spectra for the element hydrogen.

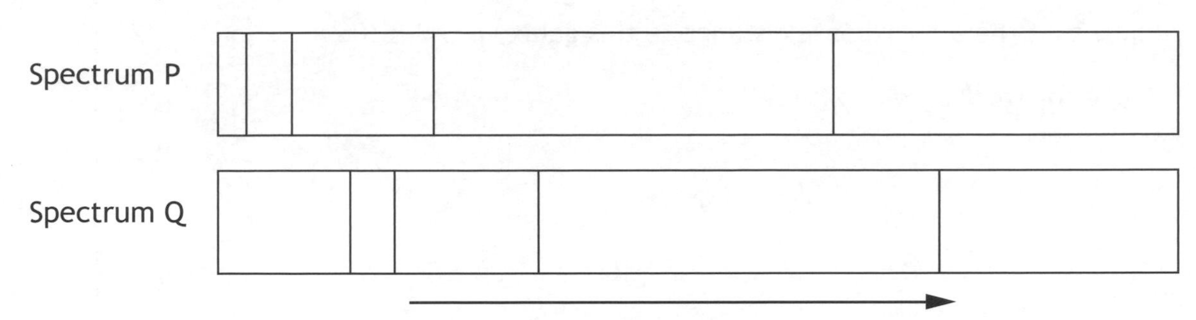

Spectrum P is from a laboratory source.

Spectrum Q shows the equivalent lines from a distant star as observed on the Earth.

(i) Explain why spectrum Q is redshifted. **2**

(ii) One of the lines in spectrum P has a wavelength of 656 nm. The equivalent line in spectrum Q is measured to have a wavelength of 676 nm.

Calculate the recessional velocity of the star. **5**

Space for working and answer

MARKS DO NOT WRITE IN THIS MARGIN

6. (continued)

(b) The recessional velocity of a distant galaxy is $1{\cdot}2 \times 10^7\,\mathrm{m\,s^{-1}}$.

Show that the approximate distance to this galaxy is $5{\cdot}2 \times 10^{24}\,\mathrm{m}$. **2**

Space for working and answer

(c) A student explains the expansion of the Universe using an "expanding balloon model".

The student draws "galaxies" on a balloon and then inflates it.

Using your knowledge of physics, comment on the suitability of this model. **3**

MARKS DO NOT WRITE IN THIS MARGIN

7. Protons and neutrons are composed of combinations of up and down quarks. Up quarks have a charge of $+\frac{2}{3}e$ while down quarks have a charge of $-\frac{1}{3}e$.

(a) (i) Determine the combination of up and down quarks that makes up:

(A) a proton; **1**

(B) a neutron. **1**

(ii) Name the boson that is the mediating particle for the strong force. **1**

(b) A neutron decays into a proton, an electron and an antineutrino.

$$^{1}_{0}\mathrm{n} \rightarrow {}^{1}_{1}\mathrm{p} + {}^{0}_{-1}e + \bar{\nu}$$

Name of this type of decay. **1**

MARKS | DO NOT WRITE IN THIS MARGIN

8. A linear accelerator is used to accelerate protons.

The accelerator consists of hollow metal tubes placed in a vacuum.

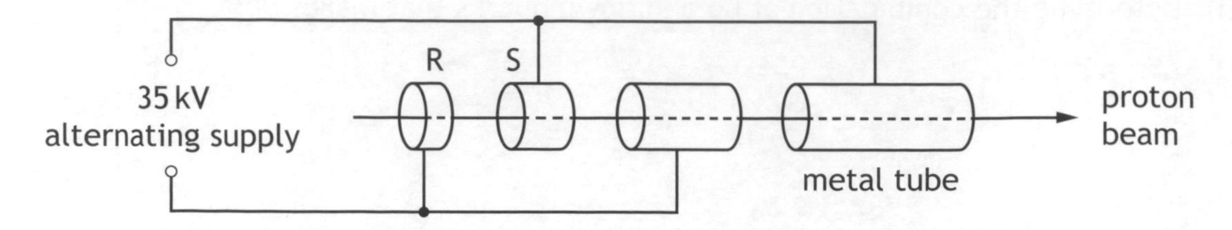

The diagram shows the path of protons through the accelerator.

Protons are accelerated across the gaps between the tubes by a potential difference of 35 kV.

(a) The protons are travelling at $1{\cdot}2 \times 10^{6}\ \mathrm{m\,s^{-1}}$ at point **R**.

(i) Show that the work done on a proton as it accelerates from **R** to **S** is $5{\cdot}6 \times 10^{-15}$ J. **2**

Space for working and answer

(ii) Calculate the speed of the proton as it reaches **S**. **5**

Space for working and answer

(b) Suggest one reason why the lengths of the tubes increase along the accelerator. **1**

MARKS DO NOT WRITE IN THIS MARGIN

9. (a) The following statement represents a fusion reaction.

$$4\,^{1}_{1}H \rightarrow \,^{4}_{2}He + 2\,^{0}_{1}e^{+}$$

The masses of the particles involved in the reaction are shown in the table.

Particle	*Mass* (kg)
$^{1}_{1}H$	$1{\cdot}673 \times 10^{-27}$
$^{4}_{2}He$	$6{\cdot}646 \times 10^{-27}$
$^{0}_{1}e$	negligible

(i) Calculate the energy released in this reaction. **4**

Space for working and answer

(ii) Calculate the energy released when 0·20 kg of hydrogen is converted to helium by this reaction. **3**

Space for working and answer

MARKS | DO NOT WRITE IN THIS MARGIN

9. (a) (continued)

(iii) Fusion reactors are being developed that use this type of reaction as an energy source.

Explain why this type of fusion reaction is hard to sustain in these reactors. **1**

(b) A nucleus of radium-224 decays to radon by emitting an alpha particle.

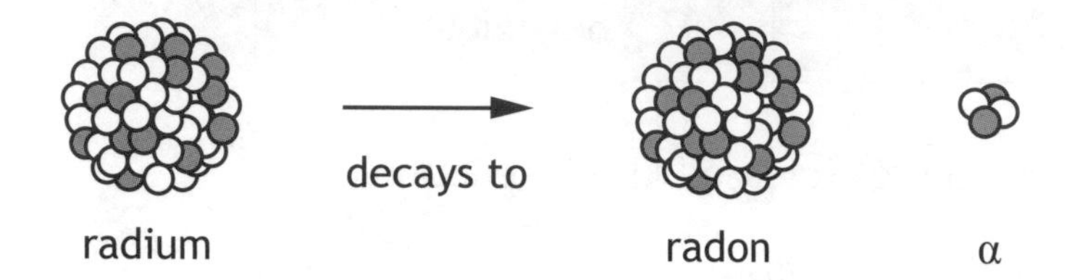

The masses of the particles involved in the decay are shown in the table.

Particle	*Mass* (kg)
radium-224	$3{\cdot}720 \times 10^{-25}$
radon-220	$3{\cdot}653 \times 10^{-25}$
alpha	$6{\cdot}645 \times 10^{-27}$

Before the decay the radium-224 nucleus is at rest.

After the decay the alpha particle moves off with a velocity of $1{\cdot}460 \times 10^7\,m\,s^{-1}$.

Calculate the velocity of the radon-220 nucleus after the decay. **3**

Space for working and answer

MARKS DO NOT WRITE IN THIS MARGIN

10. The diagram shows equipment used to investigate the photoelectric effect.

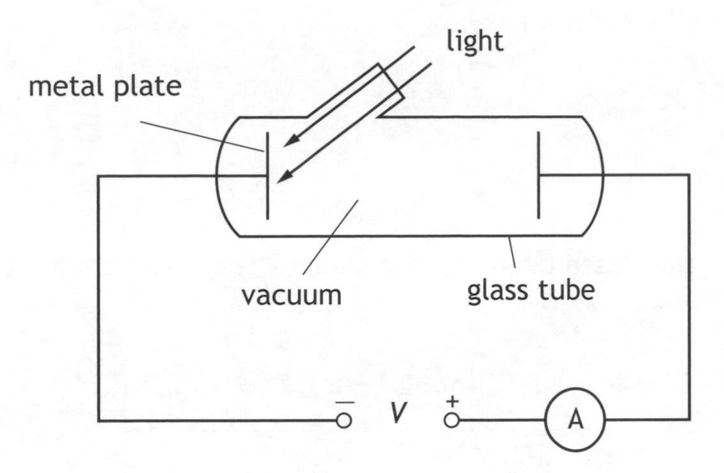

(a) When blue light is shone on the metal plate there is a current in the circuit. When blue light is replaced by red light there is no current.

Explain why this happens. **2**

(b) The blue light has a frequency of $7{\cdot}0 \times 10^{14}$ Hz.

The work function for the metal plate is $2{\cdot}0 \times 10^{-19}$ J.

Calculate the maximum kinetic energy of the electrons emitted from the plate by this light. **3**

Space for working and answer

MARKS | DO NOT WRITE IN THIS MARGIN

11. A helium-neon laser produces a beam of coherent red light.

(a) State what is meant by *coherent light*. **1**

(b) A student directs this laser beam onto a double slit arrangement as shown in the diagram.

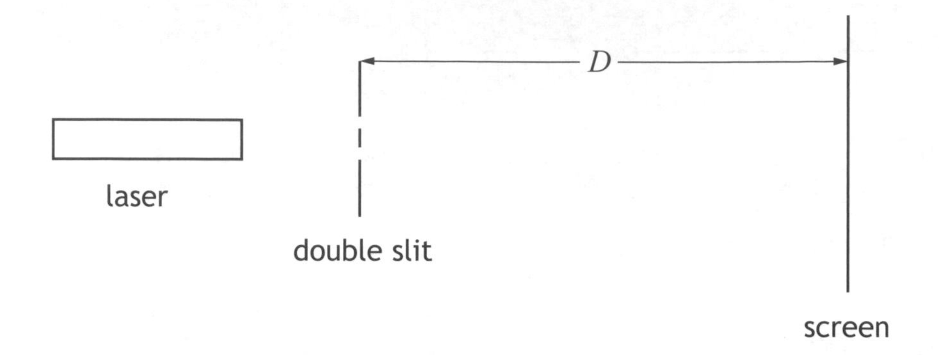

A pattern of bright red fringes is observed on the screen.

(i) Explain, in terms of waves, why bright red fringes are produced. **1**

MARKS DO NOT WRITE IN THIS MARGIN

11. (b) (continued)

(ii) The average separation, Δx, between adjacent fringes is given by the relationship

$$\Delta x = \frac{\lambda D}{d}$$

where: λ is the wavelength of the light
D is the distance between the double slit and the screen
d is the distance between the two slits

The diagram shows the value measured by the student of the distance between a series of fringes and the uncertainty in this measurement.

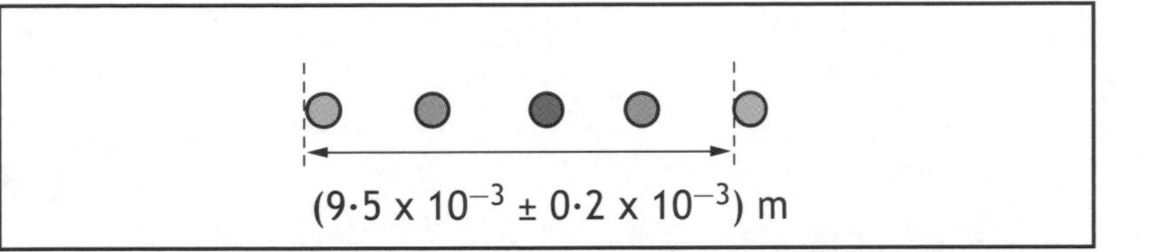

The student measures the distance, D, between the double slit and the screen as $(0{\cdot}750 \pm 0{\cdot}001)$ m.

Calculate the best estimate of the distance between the two slits.

An uncertainty in the calculated value is not required. **4**

Space for working and answer

MARKS DO NOT WRITE IN THIS MARGIN

11. (b) (continued)

(iii) The student wishes to determine more precisely the value of the distance between the two slits *d*.

Show, by calculation, which of the student's measurements should be taken more precisely in order to achieve this.

You must indicate clearly which measurement you have identified. **3**

Space for working and answer

(iv) The helium-neon laser is replaced by a laser emitting green light. No other changes are made to the experimental set-up.

Explain the effect this change has on the separation of the fringes observed on the screen. **2**

MARKS DO NOT WRITE IN THIS MARGIN

12. A student is investigating the refractive index of a Perspex block for red light.

The student directs a ray of red light towards a semicircular Perspex block as shown.

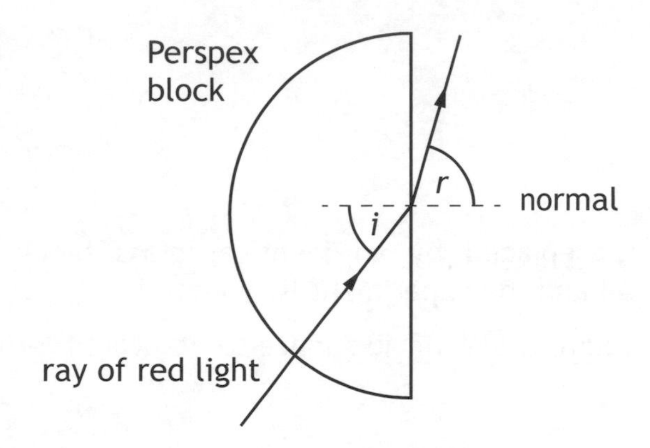

The angle of incidence *i* is then varied and the angle of refraction *r* is measured using a protractor.

The following results are obtained.

i (°)	*r* (°)	*sin i*	*sin r*
10	16	0·17	0·28
15	25	0·26	0·42
20	32	0·34	0·53
25	37	0·42	0·60
30	53	0·50	0·80

(a) (i) Using square ruled paper, draw a graph to show how sin *r* varies with sin *i*. **3**

(ii) Use the graph to determine the refractive index of the Perspex for this light. **2**

Space for working and answer

MARKS | DO NOT WRITE IN THIS MARGIN

12. (a) (continued)

(iii) Suggest **two** ways in which the experimental procedure could be improved to obtain a more accurate value for the refractive index. **2**

(b) The Perspex block is replaced by an identical glass block with a refractive index of 1·54 and the experiment is repeated.

Determine the maximum angle of incidence that would produce a refracted ray. **3**

Space for working and answer

MARKS | DO NOT WRITE IN THIS MARGIN

13. A 200 μF capacitor is charged using the circuit shown.

The 12 V battery has negligible internal resistance.

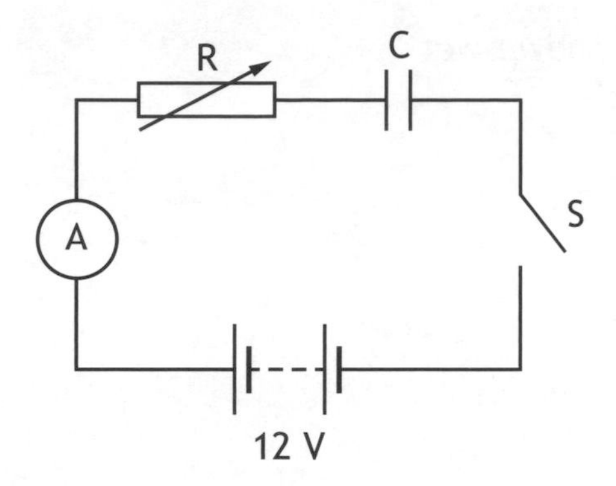

The capacitor is initially uncharged.

The switch S is closed. The charging current is kept constant at 30 μA by adjusting the resistance of the variable resistor, R.

(a) Calculate the resistance of the variable resistor R just after the switch is closed. **3**

Space for working and answer

(b) (i) Calculate the charge on the capacitor 30 s after the switch S is closed. **3**

Space for working and answer

MARKS DO NOT WRITE IN THIS MARGIN

13. (b) (continued)

(ii) Calculate the potential difference across R at this time. **4**

Space for working and answer

MARKS DO NOT WRITE IN THIS MARGIN

14. The electrical conductivity of solids can be explained by band theory.

The diagrams below show the distributions of the valence and conduction bands of materials classified as *conductors*, *insulators* and *semiconductors*.

Shaded areas represent bands occupied by electrons.

The band gap is also indicated.

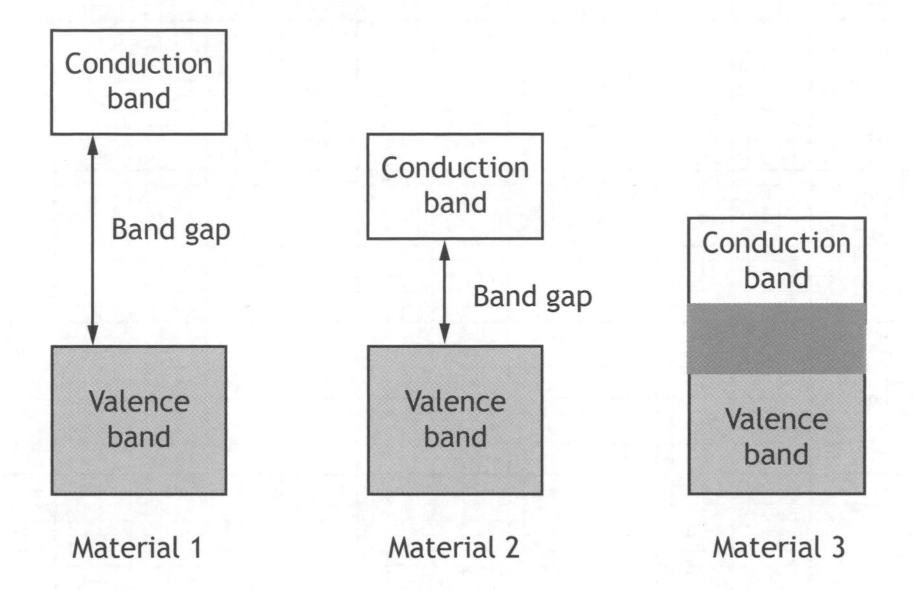

(a) State which material is a semiconductor. **1**

(b) A sample of pure semiconductor is heated. Use band theory to explain what happens to the resistance of the sample as it is heated. **2**

[END OF SPECIMEN QUESTION PAPER]

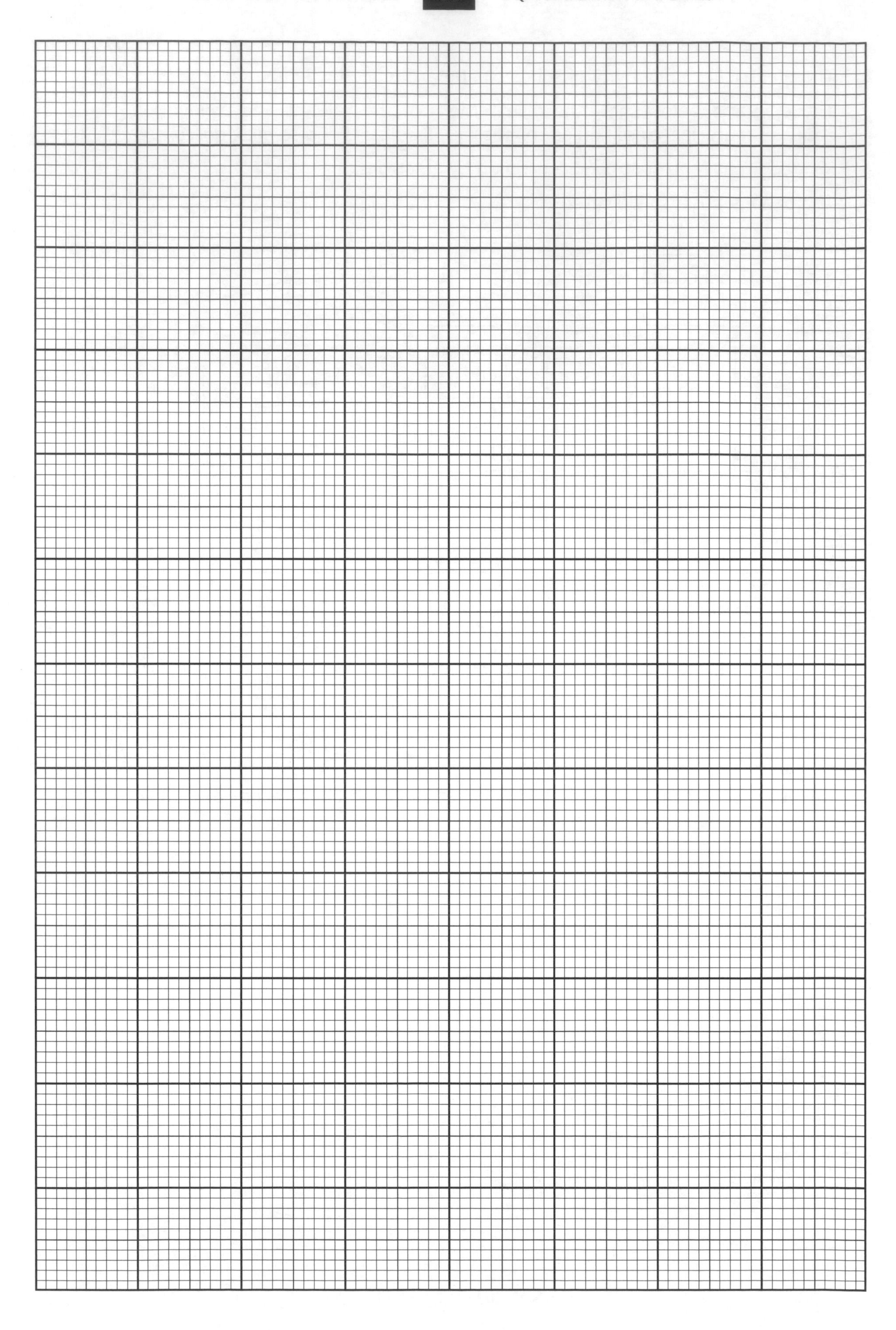

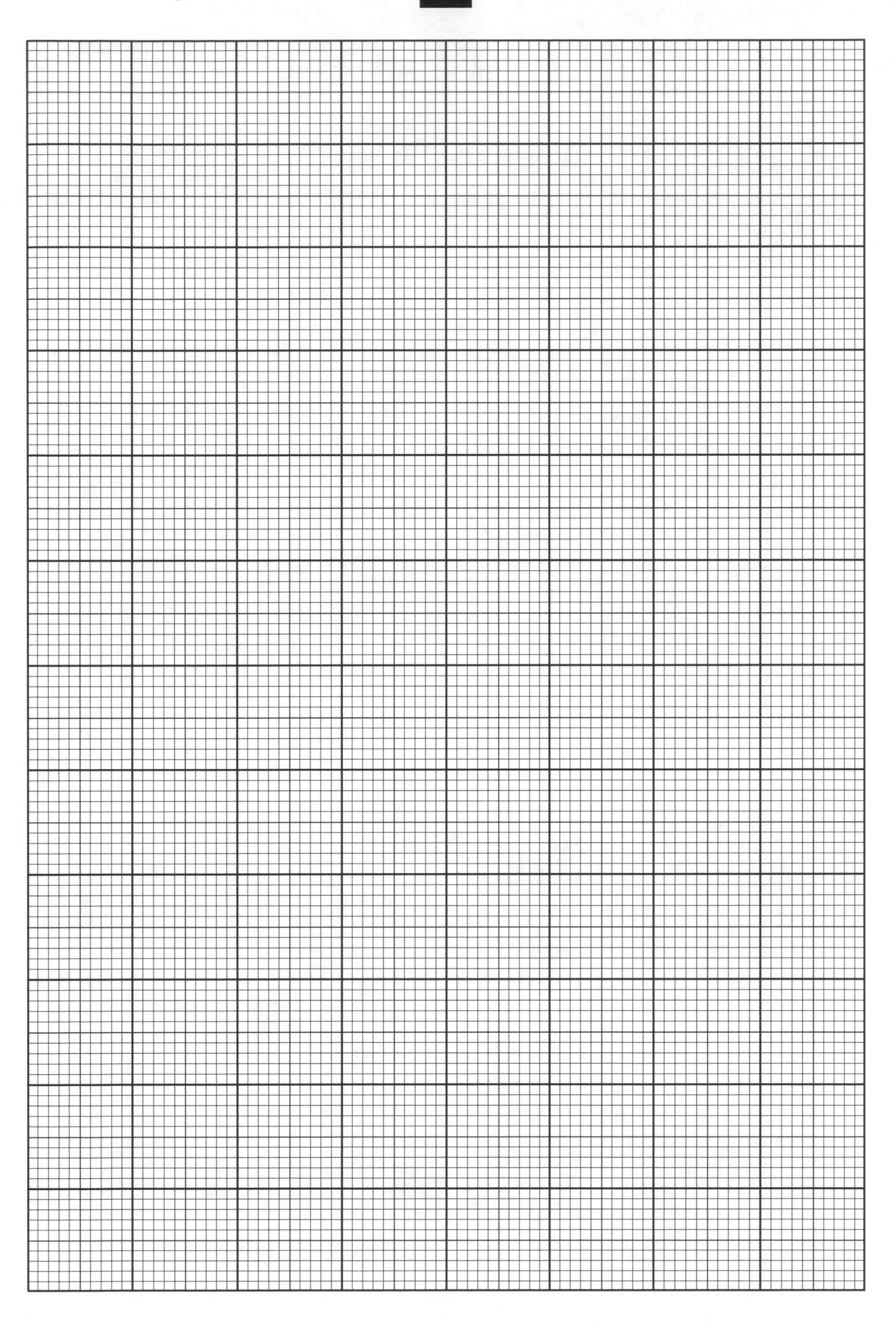

MARKS DO NOT WRITE IN THIS MARGIN

ADDITIONAL SPACE FOR ANSWERS AND ROUGH WORK

MARKS DO NOT WRITE IN THIS MARGIN

ADDITIONAL SPACE FOR ANSWERS AND ROUGH WORK

HIGHER FOR CfE

Model Paper 1

Whilst this Model Paper has been specially commissioned by Hodder Gibson for use as practice for the Higher (for Curriculum for Excellence) exams, the key reference documents remain the SQA Specimen Paper 2014 and SQA Past Paper 2015.

National
Qualifications
MODEL PAPER 1

Physics
Section 1—Questions

Duration — 2 hours and 30 minutes

Instructions for the completion of Section 1 are given on *Page two* of your question and answer booklet.

Record your answers on the answer grid on *Page three* of your question and answer booklet.

Reference may be made to the Data Sheet on *Page two* of this booklet and to the Relationships Sheet.

Before leaving the examination room you must give your question and answer booklet to the Invigilator; if you do not, you may lose all the marks for this paper.

DATA SHEET

COMMON PHYSICAL QUANTITIES

Quantity	*Symbol*	*Value*	*Quantity*	*Symbol*	*Value*
Speed of light in vacuum	c	$3{\cdot}00 \times 10^{8}$ m s^{-1}	Planck's constant	h	$6{\cdot}63 \times 10^{-34}$ J s
Magnitude of the charge on an electron	e	$1{\cdot}60 \times 10^{-19}$ C	Mass of electron	m_e	$9{\cdot}11 \times 10^{-31}$ kg
Universal Constant of Gravitation	G	$6{\cdot}67 \times 10^{-11}$ m^{3} kg^{-1} s^{-2}	Mass of neutron	m_n	$1{\cdot}675 \times 10^{-27}$ kg
Gravitational acceleration on Earth	g	$9{\cdot}8$ m s^{-2}	Mass of proton	m_p	$1{\cdot}673 \times 10^{-27}$ kg
Hubble's constant	H_0	$2{\cdot}3 \times 10^{-18}$ s^{-1}			

REFRACTIVE INDICES

The refractive indices refer to sodium light of wavelength 589 nm and to substances at a temperature of 273 K.

Substance	*Refractive index*	*Substance*	*Refractive index*
Diamond	2·42	Water	1·33
Crown glass	1·50	Air	1·00

SPECTRAL LINES

Element	*Wavelength*/nm	*Colour*
Hydrogen	656	Red
	486	Blue-green
	434	Blue-violet
	410	Violet
	397	Ultraviolet
	389	Ultraviolet
Sodium	589	Yellow

Element	*Wavelength*/nm	*Colour*
Cadmium	644	Red
	509	Green
	480	Blue

Lasers

Element	*Wavelength*/nm	*Colour*
Carbon dioxide	9550 } 10590 }	Infrared
Helium-neon	633	Red

PROPERTIES OF SELECTED MATERIALS

Substance	*Density*/kg m^{-3}	*Melting Point*/K	*Boiling Point*/K
Aluminium	$2{\cdot}70 \times 10^{3}$	933	2623
Copper	$8{\cdot}96 \times 10^{3}$	1357	2853
Ice	$9{\cdot}20 \times 10^{2}$	273	....
Sea Water	$1{\cdot}02 \times 10^{3}$	264	377
Water	$1{\cdot}00 \times 10^{3}$	273	373
Air	1·29	....	....
Hydrogen	$9{\cdot}0 \times 10^{-2}$	14	20

The gas densities refer to a temperature of 273 K and a pressure of $1{\cdot}01 \times 10^{5}$ Pa.

SECTION 1 — 20 marks

Attempt ALL questions

1. A trolley travels along a straight track.

The graph shows how the velocity v of the trolley varies with time t.

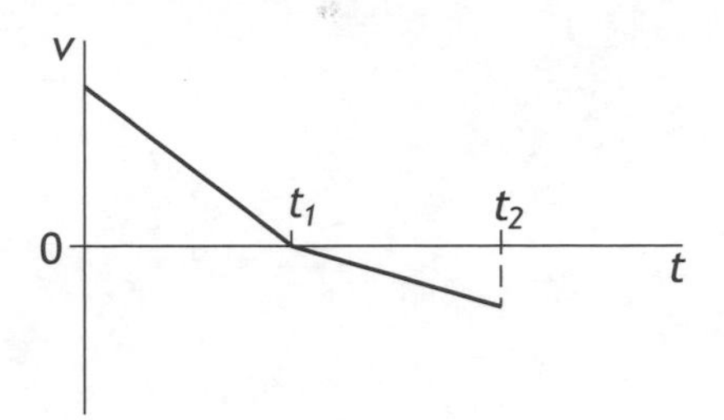

Which graph shows how the acceleraton a of the trolley varies with time t?

A

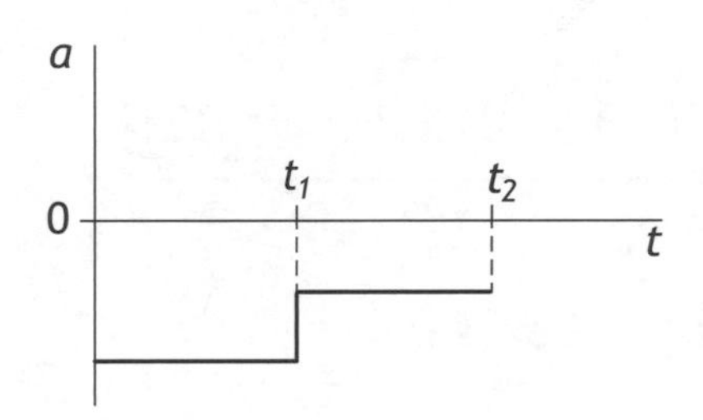

B

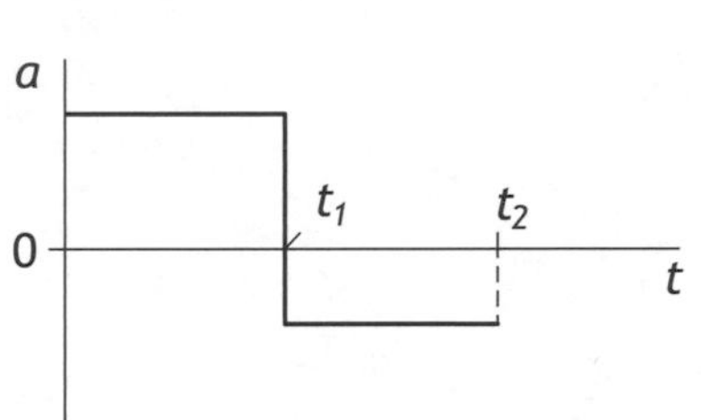

C

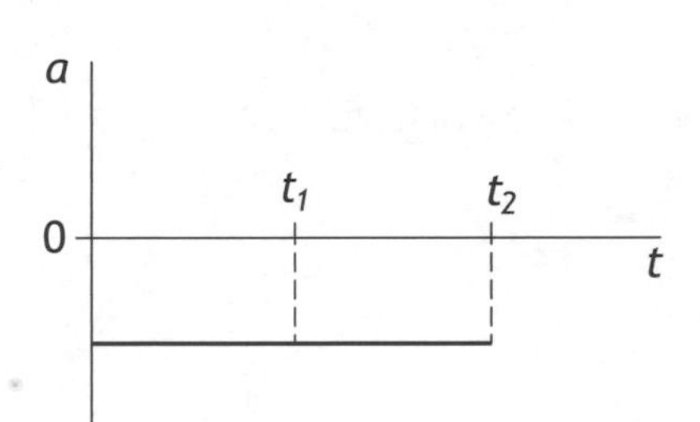

D

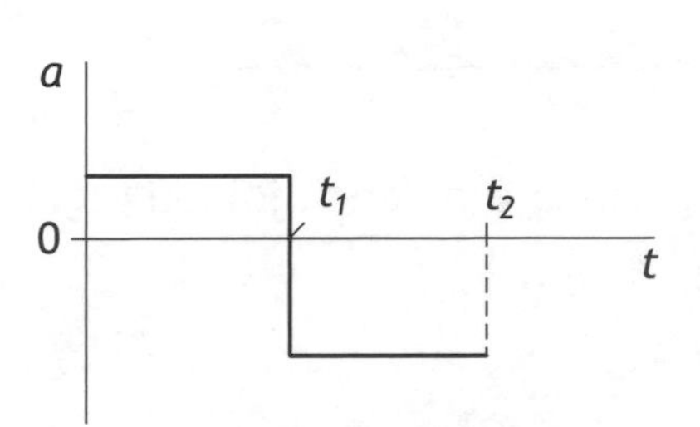

E

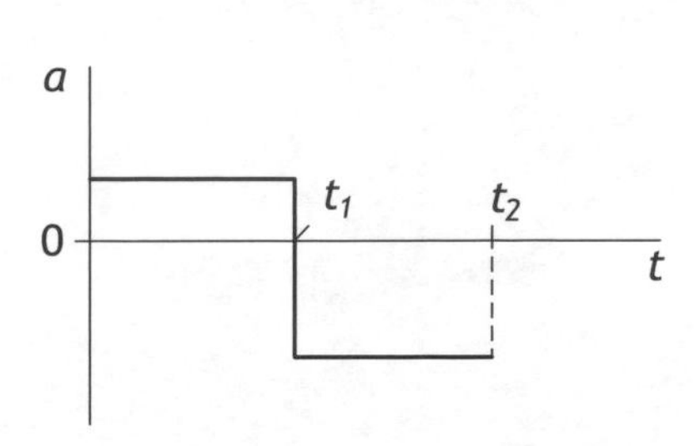

2. A helicopter is **descending** vertically at a constant speed of 3·0 m s^{-1}. A sandbag is released from the helicopter. The sandbag hits the ground 5·0 s later.

   What is the height of the helicopter above the ground at the time the sandbag was released?

   A 15·0 m

   B 49·0 m

   C 107·5 m

   D 122·5 m

   E 137·5 m

3. Two boxes on a frictionless horizontal surface are joined together by a string. A constant horizontal force of 12 N is applied as shown.

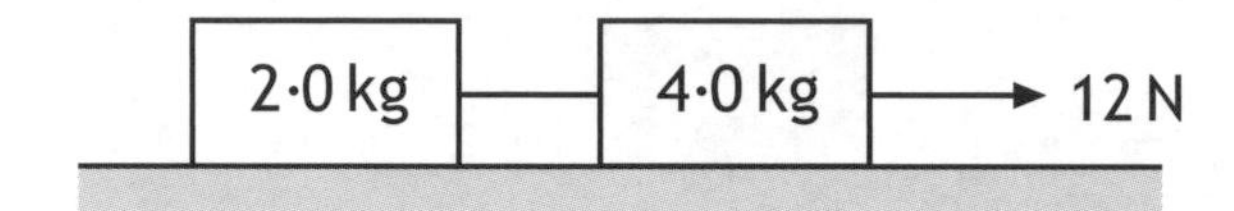

   The tension in the string joining the two boxes is

   A 2·0 N

   B 4·0 N

   C 6·0 N

   D 8·0 N

   E 12 N.

4. The diagram shows the masses and velocities of two trolleys just before they collide on a level bench.

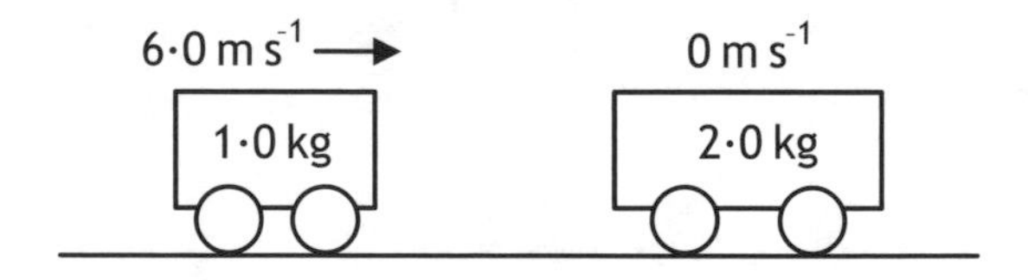

   After the collision, the trolleys move along the bench joined together.

   How much kinetic energy is lost in this collision?

   A 0 J

   B 6·0 J

   C 12 J

   D 18 J

   E 24 J

5. The acceleration at the surface of the earth is taken as 9.8 ms^{-2}. The acceleration at 4 earth radii from the **surface** of the earth will be:

A 2.5 ms^{-2}

B 0.39 ms^{-2}

C 4.9 ms^{-2}

D 0.61 ms^{-2}

E 0.

6. A police car travelling at constant speed, with its siren on, approaches a pedestrian.

Which statement correctly describes the frequency of the sound heard by the pedestrian?

A The frequency increases as the car approaches and decreases as the car passes.

B The frequency is constant as the car approaches and decreases as the car passes.

C The frequency is constant as the car approaches and increases as the car passes.

D The frequency increases as the car approaches and increases as the car passes.

E The frequency decreases as the car approaches and increases as the car passes.

7. Which of the following statements did Einstein postulate in his Theory of Special Relativity.

I When two observers are moving at constant speeds relative to one another, they will observe the same laws of physics.

II Travelling near the speed of light will cause time to slow down

III The speed of light (in a vacuum) is the same for all observers, regardless of their motion relative to the light source.

A I only

B II only

C I and II

D I and III

E I, II and III.

8. An example of a lepton is a:

A quark
B neutron
C antiproton
D muon
E photon

9. Waves from coherent sources, S_1 and S_2, produce an interference pattern. Maxima of intensity are detected at the positions shown below.

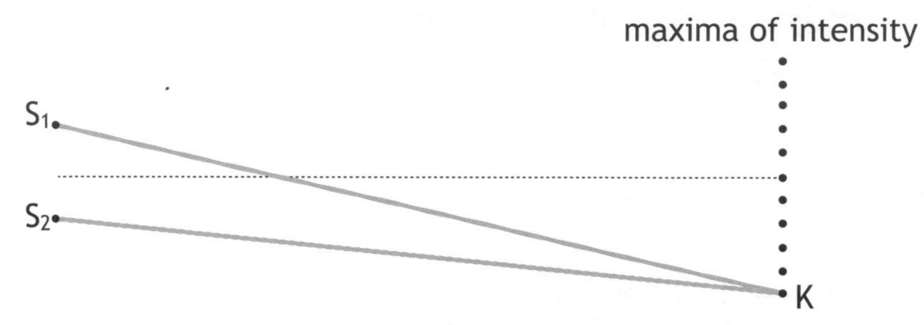

The path difference $S_1K - S_2K$ is 154 mm. The wavelength of the waves is

A 15.4 mm
B 25.7 mm
C 28.0 mm
D 30.8 mm
E 34.2 mm

10. A ray of monochromatic light passes into a glass block as shown.

The refractive index of the glass for this light is

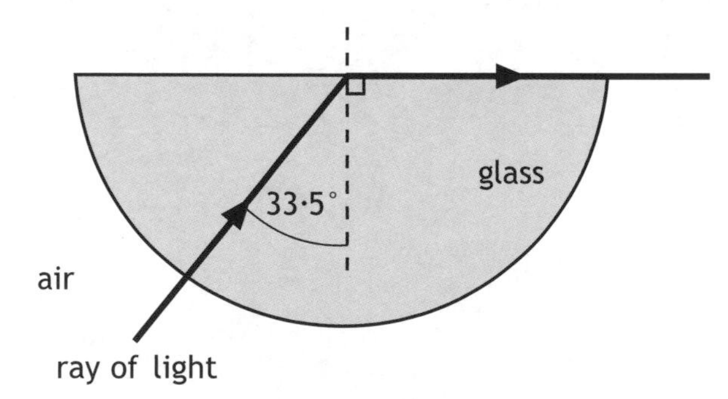

A 0.03
B 0.55
C 0.87
D 1.20
E 1.81.

11. The diagram represents a ray of light passing from air into liquid.

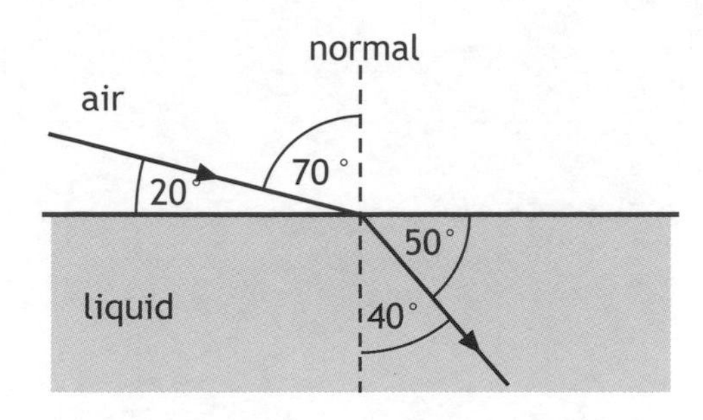

The refractive index of this liquid, relative to air, is

A $\frac{\sin 20°}{\sin 40°}$

B $\frac{\sin 40°}{\sin 70°}$

C $\frac{\sin 50°}{\sin 20°}$

D $\frac{\sin 70°}{\sin 40°}$

E $\frac{\sin 90°}{\sin 40°}$.

12. When light of frequency f is shone on to a certain metal, photoelectrons are ejected with a maximum velocity v and kinetic energy E_k.

Light of the same frequency but twice the irradiance is shone on to the same surface.

Which of the following statements is/are correct?

I Twice as many electrons are ejected per second.

II The speed of the fastest electron is 2 v.

III The kinetic energy of the fastest electron is now 2 E_k.

A I only

B II only

C III only

D I and II only

E I, II and III

13. α particles, β particles and neutrons are fired at 90° through a magnetic field as shown.

α, β and n →

X X X X
X X X X
X X X X
X X X X

Magnetic field into the paper

Which entry below correctly describes the path of the particles?

	Deflected upwards	Deflected downwards	Undeflected
A	β	$\propto$	n
B	β	n	$\propto$
C	n	β	$\propto$
D	$\propto$	n	β
E	$\propto$	β	n

14. The irradiance of light from a point source is 20 W m^{-2} at a distance of 5·0 m from the source.

What is the irradiance of the light at a distance of 25 m from the source?

A 0·032 W m^{-2}

B 0·80 W m^{-2}

C 4·0 W m^{-2}

D 100 W m^{-2}

E 500 W m^{-2}

15. A battery of e.m.f. 24 V and negligible internal resistance is connected as shown.

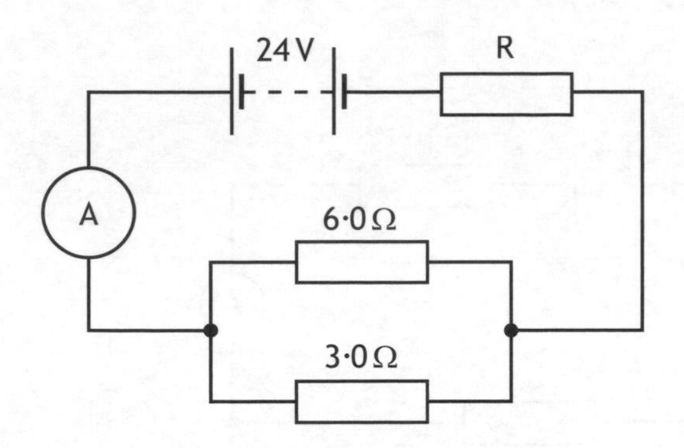

The reading on the ammeter is 2.0A.

The resistance of R is

A 3.0 Ω

B 4.0 Ω

C 10 Ω

D 12 Ω

E 18 Ω.

16. Three resistors are connected as shown.

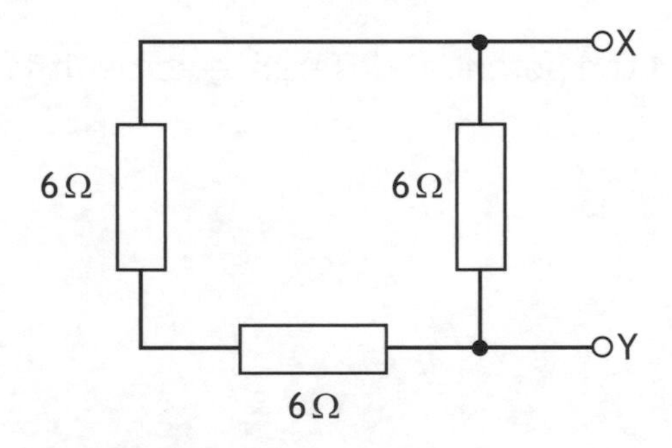

The total resistance between X and Y is

A 2 Ω

B 4 Ω

C 6 Ω

D 9 Ω

E 18 Ω.

17. A battery of e.m.f. 12 V and internal resistance 3.0Ω is connected in a circuit as shown.

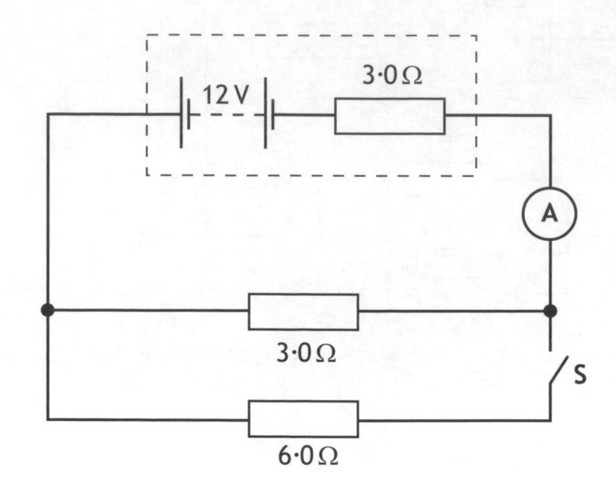

When switch S is closed the ammeter reading changes from

A 2·0 A to 1·0 A

B 2·0 A to 2·4 A

C 2·0 A to 10 A

D 4·0 A to 1·3 A

E 4·0 A to 6·0 A.

18. A 25·0 μF capacitor is charged until the potential difference across it is 500 V.

The charge stored in the capacitor is

A $5{\cdot}00 \times 10^{-8}$ C

B $2{\cdot}00 \times 10^{-5}$ C

C $1{\cdot}25 \times 10^{-2}$ C

D $1{\cdot}25 \times 10^{4}$ C

E $2{\cdot}00 \times 10^{7}$ C.

19. A circuit is set up as shown.

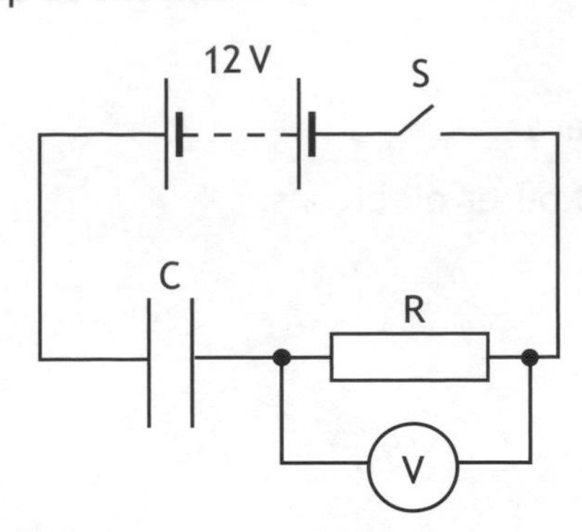

The capacitor is initially uncharged. Switch S is now closed. Which graph shows how the potential difference, V, across R, varies with time, t?

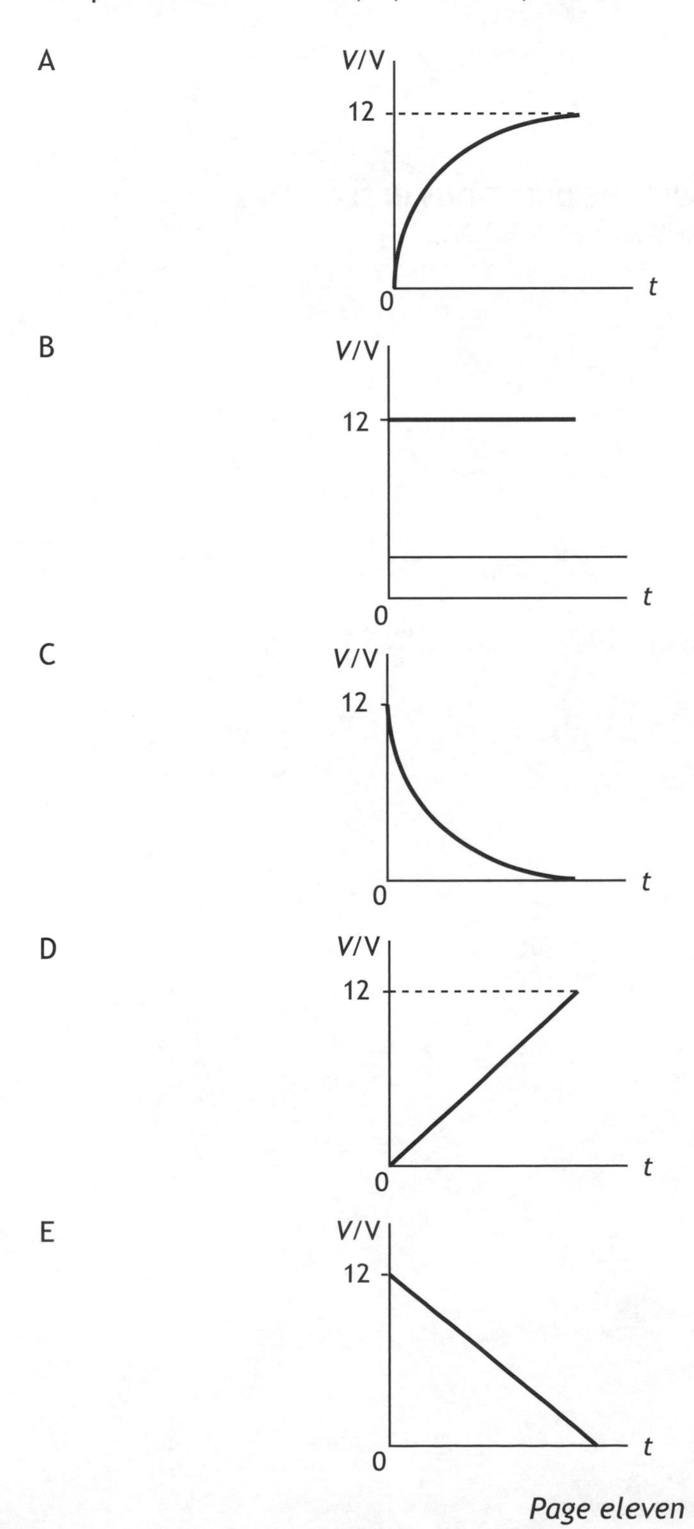

20. A student writes the following statements about p-type semiconductor material.

I Most charge carriers are positive.

II The p-type material has a positive charge.

III Impurity atoms in the material have 3 outer electrons.

Which of these statements is/are true?

A I only

B II only

C I and II only

D I and III only

E I, II and III

[END OF SECTION 1. NOW ATTEMPT THE QUESTIONS IN SECTION 2 OF YOUR QUESTION AND ANSWER BOOKLET]

National
Qualifications
MODEL PAPER 1

Physics
Relationships Sheet

Date — Not applicable

Relationships required for Physics Higher

$d = \bar{v}t$

$s = \bar{v}t$

$v = u + at$

$s = ut + \frac{1}{2}at^2$

$v^2 = u^2 + 2as$

$s = \frac{1}{2}(u + v)t$

$W = mg$

$F = ma$

$E_W = Fd$

$E_p = mgh$

$E_k = \frac{1}{2}mv^2$

$P = \frac{E}{t}$

$p = mv$

$Ft = mv - mu$

$F = G\frac{m_1 m_2}{r^2}$

$t' = \frac{t}{\sqrt{1-\left(\frac{v}{c}\right)^2}}$

$l' = l\sqrt{1-\left(\frac{v}{c}\right)^2}$

$f_o = f_s\left(\frac{v}{v \pm v_s}\right)$

$z = \frac{\lambda_{observed} - \lambda_{rest}}{\lambda_{rest}}$

$z = \frac{v}{c}$

$v = H_0 d$

$W = QV$

$E = mc^2$

$E = hf$

$E_k = hf - hf_0$

$E_2 - E_1 = hf$

$T = \frac{1}{f}$

$v = f\lambda$

$d \sin\theta = m\lambda$

$n = \frac{\sin\theta_1}{\sin\theta_2}$

$\frac{\sin\theta_1}{\sin\theta_2} = \frac{\lambda_1}{\lambda_2} = \frac{v_1}{v_2}$

$\sin\theta_c = \frac{1}{n}$

$I = \frac{k}{d^2}$

$I = \frac{P}{A}$

$V_{peak} = \sqrt{2}V_{rms}$

$I_{peak} = \sqrt{2}I_{rms}$

$Q = It$

$V = IR$

$P = IV = I^2R = \frac{V^2}{R}$

$R_T = R_1 + R_2 + \ldots.$

$\frac{1}{R_T} = \frac{1}{R_1} + \frac{1}{R_2} + \ldots.$

$E = V + Ir$

$V_1 = \left(\frac{R_1}{R_1 + R_2}\right)V_s$

$\frac{V_1}{V_2} = \frac{R_1}{R_2}$

$C = \frac{Q}{V}$

$E = \frac{1}{2}QV = \frac{1}{2}CV^2 = \frac{1}{2}\frac{Q^2}{C}$

path difference = $m\lambda$ or $\left(m + \frac{1}{2}\right)\lambda$ where $m = 0, 1, 2 \ldots$

random uncertainty = $\frac{\text{max. value} - \text{min. value}}{\text{number of values}}$

Additional Relationships

Circle

circumference = $2\pi r$

area = πr^2

Sphere

area = $4\pi r^2$

volume = $\frac{4}{3}\pi r^3$

Trigonometry

$\sin\Theta = \frac{\text{opposite}}{\text{hypotenuse}}$

$\cos\Theta = \frac{\text{adjacent}}{\text{hypotenuse}}$

$\tan\Theta = \frac{\text{opposite}}{\text{adjacent}}$

$\sin^2\Theta + \cos^2\Theta = 1$

Electron Arrangements of Elements

Key

Atomic number Symbol Electron arrangement Name

Group 1 (1)	Group 2 (2)
1 **H** 1 Hydrogen	
3 **Li** 2,1 Lithium	4 **Be** 2,2 Beryllium
11 **Na** 2,8,1 Sodium	12 **Mg** 2,8,2 Magnesium
19 **K** 2,8,8,1 Potassium	20 **Ca** 2,8,8,2 Calcium
37 **Rb** 2,8,18,8,1 Rubidium	38 **Sr** 2,8,18,8,2 Strontium
55 **Cs** 2,8,18,18, 8,1 Caesium	56 **Ba** 2,8,18,18, 8,2 Barium
87 **Fr** 2,8,18,32, 18,8,1 Francium	88 **Ra** 2,8,18,32, 18,8,2 Radium

Transition Elements

(3)	(4)	(5)	(6)	(7)	(8)	(9)	(10)	(11)	(12)
21 **Sc** 2,8,9,2 Scandium	22 **Ti** 2,8,10,2 Titanium	23 **V** 2,8,11,2 Vanadium	24 **Cr** 2,8,13,1 Chromium	25 **Mn** 2,8,13,2 Manganese	26 **Fe** 2,8,14,2 Iron	27 **Co** 2,8,15,2 Cobalt	28 **Ni** 2,8,16,2 Nickel	29 **Cu** 2,8,18,1 Copper	30 **Zn** 2,8,18,2 Zinc
39 **Y** 2,8,18,9,2 Yttrium	40 **Zr** 2,8,18, 10,2 Zirconium	41 **Nb** 2,8,18, 12,1 Niobium	42 **Mo** 2,8,18,13, 1 Molybdenum	43 **Tc** 2,8,18,13, 2 Technetium	44 **Ru** 2,8,18,15, 1 Ruthenium	45 **Rh** 2,8,18,16, 1 Rhodium	46 **Pd** 2,8,18, 18,0 Palladium	47 **Ag** 2,8,18, 18,1 Silver	48 **Cd** 2,8,18, 18,2 Cadmium
57 **La** 2,8,18,18, 9,2 Lanthanum	72 **Hf** 2,8,18,32, 10,2 Hafnium	73 **Ta** 2,8,18, 32,11,2 Tantalum	74 **W** 2,8,18,32, 12,2 Tungsten	75 **Re** 2,8,18,32, 13,2 Rhenium	76 **Os** 2,8,18,32, 14,2 Osmium	77 **Ir** 2,8,18,32, 15,2 Iridium	78 **Pt** 2,8,18,32, 17,1 Platinum	79 **Au** 2,8,18, 32,18,1 Gold	80 **Hg** 2,8,18, 32,18,2 Mercury
89 **Ac** 2,8,18,32, 18,9,2 Actinium	104 **Rf** 2,8,18,32, 32,10,2 Rutherfordium	105 **Db** 2,8,18,32, 32,11,2 Dubnium	106 **Sg** 2,8,18,32, 32,12,2 Seaborgium	107 **Bh** 2,8,18,32, 32,13,2 Bohrium	108 **Hs** 2,8,18,32, 32,14,2 Hassium	109 **Mt** 2,8,18,32, 32,15,2 Meitnerium	110 **Ds** 2,8,18,32, 32,17,1 Darmstadtium	111 **Rg** 2,8,18,32, 32,18,1 Roentgenium	112 **Cn** 2,8,18,32, 32,18,2 Copernicium

Group 3 (13)	Group 4 (14)	Group 5 (15)	Group 6 (16)	Group 7 (17)	Group 0 (18)
					2 **He** 2 Helium
5 **B** 2,3 Boron	6 **C** 2,4 Carbon	7 **N** 2,5 Nitrogen	8 **O** 2,6 Oxygen	9 **F** 2,7 Fluorine	10 **Ne** 2,8 Neon
13 **Al** 2,8,3 Aluminium	14 **Si** 2,8,4 Silicon	15 **P** 2,8,5 Phosphorus	16 **S** 2,8,6 Sulfur	17 **Cl** 2,8,7 Chlorine	18 **Ar** 2,8,8 Argon
31 **Ga** 2,8,18,3 Gallium	32 **Ge** 2,8,18,4 Germanium	33 **As** 2,8,18,5 Arsenic	34 **Se** 2,8,18,6 Selenium	35 **Br** 2,8,18,7 Bromine	36 **Kr** 2,8,18,8 Krypton
49 **In** 2,8,18, 18,3 Indium	50 **Sn** 2,8,18, 18,4 Tin	51 **Sb** 2,8,18, 18,5 Antimony	52 **Te** 2,8,18, 18,6 Tellurium	53 **I** 2,8,18, 18,7 Iodine	54 **Xe** 2,8,18, 18,8 Xenon
81 **Tl** 2,8,18, 32,18,3 Thallium	82 **Pb** 2,8,18, 32,18,4 Lead	83 **Bi** 2,8,18, 32,18,5 Bismuth	84 **Po** 2,8,18, 32,18,6 Polonium	85 **At** 2,8,18, 32,18,7 Astatine	86 **Rn** 2,8,18, 32,18,8 Radon

Lanthanides	57 **La** 2,8,18, 18,9,2 Lanthanum	58 **Ce** 2,8,18, 20,8,2 Cerium	59 **Pr** 2,8,18,21, 8,2 Praseodymium	60 **Nd** 2,8,18,22, 8,2 Neodymium	61 **Pm** 2,8,18,23, 8,2 Promethium	62 **Sm** 2,8,18,24, 8,2 Samarium	63 **Eu** 2,8,18,25, 8,2 Europium	64 **Gd** 2,8,18,25, 9,2 Gadolinium	65 **Tb** 2,8,18,27, 8,2 Terbium	66 **Dy** 2,8,18,28, 8,2 Dysprosium	67 **Ho** 2,8,18,29, 8,2 Holmium	68 **Er** 2,8,18,30, 8,2 Erbium	69 **Tm** 2,8,18,31, 8,2 Thulium	70 **Yb** 2,8,18,32, 8,2 Ytterbium	71 **Lu** 2,8,18,32, 9,2 Lutetium
Actinides	89 **Ac** 2,8,18,32, 18,9,2 Actinium	90 **Th** 2,8,18,32, 18,10,2 Thorium	91 **Pa** 2,8,18,32, 20,9,2 Protactinium	92 **U** 2,8,18,32, 21,9,2 Uranium	93 **Np** 2,8,18,32, 22,9,2 Neptunium	94 **Pu** 2,8,18,32, 24,8,2 Plutonium	95 **Am** 2,8,18,32, 25,8,2 Americium	96 **Cm** 2,8,18,32, 25,9,2 Curium	97 **Bk** 2,8,18,32, 27,8,2 Berkelium	98 **Cf** 2,8,18,32, 28,8,2 Californium	99 **Es** 2,8,18,32, 29,8,2 Einsteinium	100 **Fm** 2,8,18,32, 30,8,2 Fermium	101 **Md** 2,8,18,32, 31,8,2 Mendelevium	102 **No** 2,8,18,32, 32,8,2 Nobelium	103 **Lr** 2,8,18,32, 32,9,2 Lawrencium

Physics
Section 1 — Answer Grid and Section 2

Duration — 2 hours and 30 minutes

Fill in these boxes and read what is printed below.

Full name of centre

Town

Forename(s)

Surname

Number of seat

Date of birth

Day Month Year

D D M M Y Y

Scottish candidate number

Total marks — 130

SECTION 1 — 20 marks

Attempt ALL questions.

Instructions for the completion of Section 1 are given on *Page two*.

SECTION 2 — 110 marks

Attempt ALL questions.

Reference may be made to the Data Sheet on *Page two* of the question paper and to the Relationship Sheet.

Write your answers clearly in the spaces provided in this booklet. Additional space for answers and rough work is provided at the end of this booklet. If you use this space you must clearly identify the question number you are attempting. Any rough work must be written in this booklet. You should score through your rough work when you have written your final copy.

Use **blue** or **black** ink.

Care should be taken to give an appropriate number of significant figures in the final answers to calculations.

Before leaving the examination room you must give this booklet to the Invigilator; if you do not you may lose all the marks for this paper.

HODDER GIBSON
LEARN MORE

SECTION 1 — 20 marks

The questions for Section 1 are contained in the booklet Physics Section 1 – Questions.
Read these and record your answers on the answer grid on *Page three* opposite.
Do **NOT** use gel pens.

1. The answer to each question is **either** A, B, C, D or E. Decide what your answer is, then fill in the appropriate bubble (see sample question below).

2. There is **only one correct** answer to each question.

3. Any rough working should be done on the additional space for answers and rough work at the end of this booklet.

Sample Question

The energy unit measured by the electricity meter in your home is the:

A ampere

B kilowatt-hour

C watt

D coulomb

E volt.

The correct answer is **B**—kilowatt-hour. The answer **B** bubble has been clearly filled in (see below).

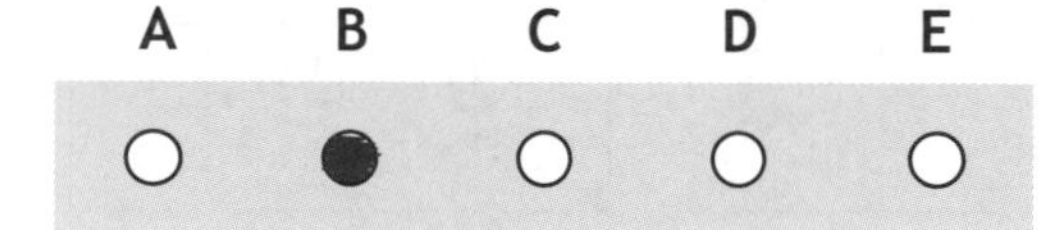

Changing an answer

If you decide to change your answer, cancel your first answer by putting a cross through it (see below) and fill in the answer you want. The answer below has been changed to **D**.

A B C D E

If you then decide to change back to an answer you have already scored out, put a tick (✓) to the **right** of the answer you want, as shown below:

A B C D E

or

A B C D E

SECTION 1 — Answer Grid

	A	B	C	D	E
1	○	○	○	○	○
2	○	○	○	○	○
3	○	○	○	○	○
4	○	○	○	○	○
5	○	○	○	○	○
6	○	○	○	○	○
7	○	○	○	○	○
8	○	○	○	○	○
9	○	○	○	○	○
10	○	○	○	○	○
11	○	○	○	○	○
12	○	○	○	○	○
13	○	○	○	○	○
14	○	○	○	○	○
15	○	○	○	○	○
16	○	○	○	○	○
17	○	○	○	○	○
18	○	○	○	○	○
19	○	○	○	○	○
20	○	○	○	○	○

MARKS DO NOT WRITE IN THIS MARGIN

SECTION 2 — 110 marks

Attempt ALL questions

1. A pupil conducts the following experiment to find the acceleration due to gravity by timing a falling ball through different distances. The ball contains a timer that starts when dropped and stops when it hits the ground.

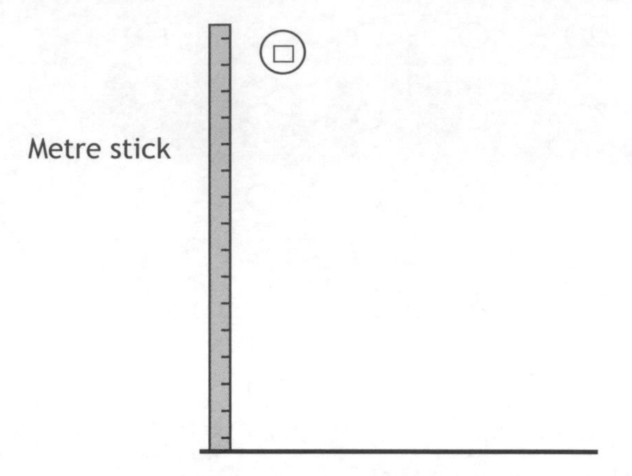

(a) Assuming the ball falls from rest, show that the height, h, travelled will be given by:

$h = 0{\cdot}5\ at^2$

where a = acceleration due to gravity

and t = the time taken to fall a height, h. **1**

MARKS DO NOT WRITE IN THIS MARGIN

1. (continued)

(b) The results of the experiment are shown below.

Complete the table: **2**

Height (m)	Times (s)	Mean time, t (s)	t^2 (s^2)
0·10	0·11, 0·14, 0·13		
0·20	0·21, 0·19, 0·17		
0·30	0·21, 0·23, 0·24		
0·40	0·25, 0·26, 0·30		
0·50	0·32, 0·28, 0·30		

(c) Plot the graph of h against t^2. **3**

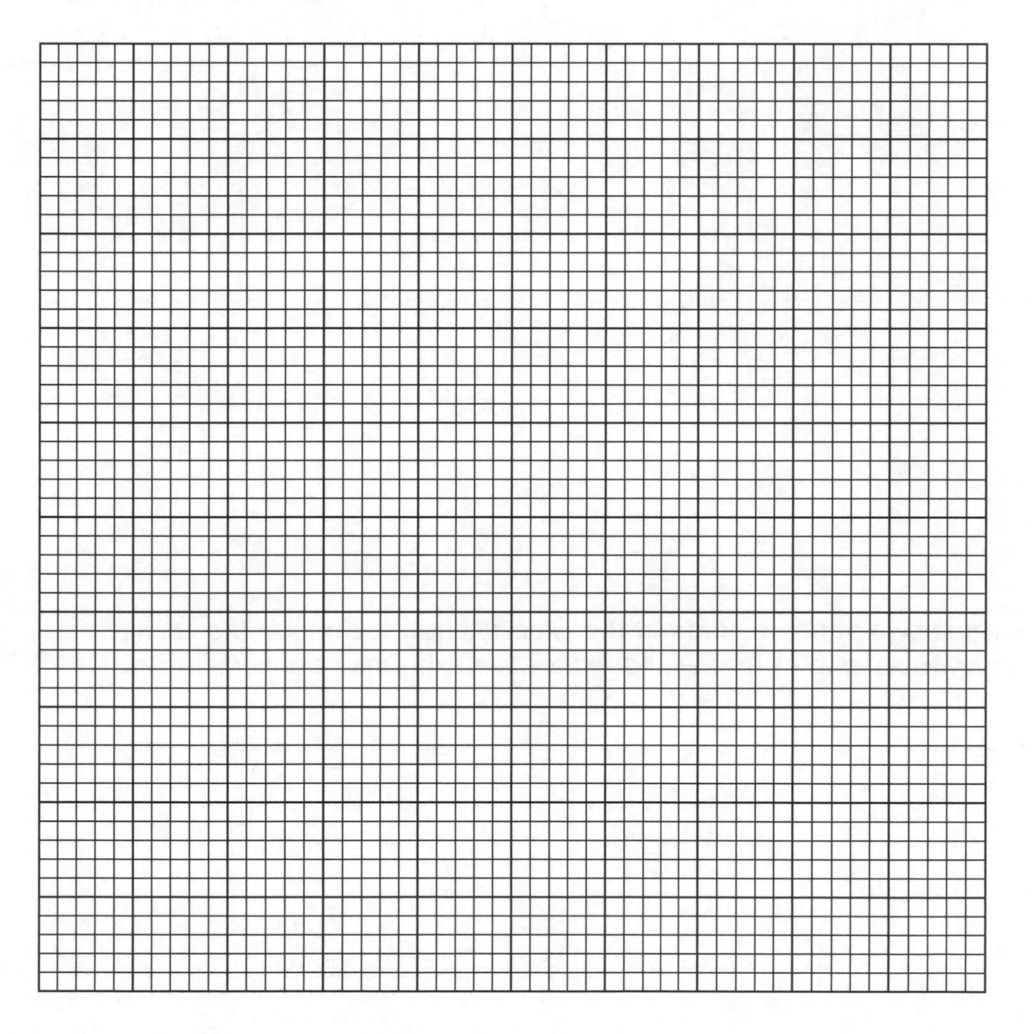

MARKS DO NOT WRITE IN THIS MARGIN

1. **(continued)**

(d) Using the graph, calculate the acceleration due to gravity. 4

(e) Suggest **two** improvements to the experimental technique (no change of apparatus possible) that would reduce the experiment uncertainties. 2

1. ____

2. ____

Total marks 12

MARKS | DO NOT WRITE IN THIS MARGIN

2. (a) The π^+ meson is an unstable particle with a lifetime of $2{\cdot}4 \times 10^{-6}$ s. The particle travels at a constant velocity relative to a laboratory and its lifetime is found to be 4.8×10^{-6} s.

Calculate its velocity relative to the laboratory. **3**

(b) What distance, measured in the lab, does the particle travel during its lifetime? **3**

Total marks 6

MARKS | DO NOT WRITE IN THIS MARGIN

3. A basketball player throws a ball with an initial velocity of $6 \cdot 5 \text{ ms}^{-1}$ at an angle of 50° to the horizontal. The ball is 2·3 m above the ground when released.

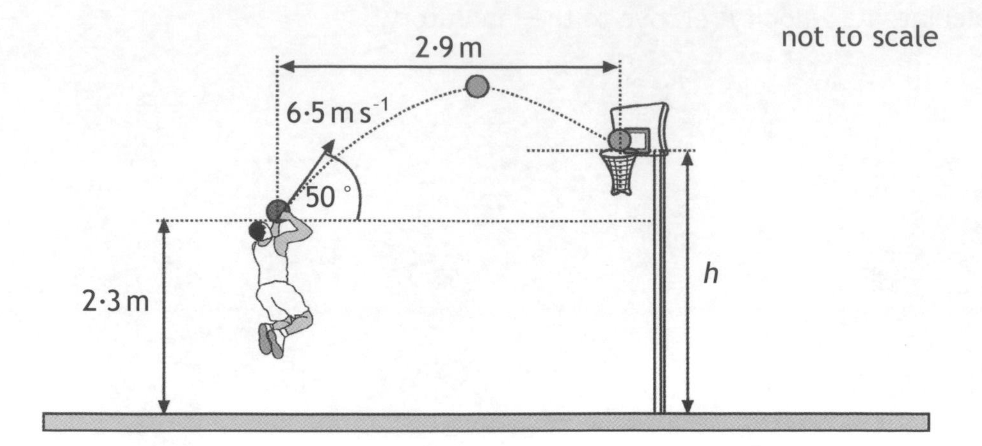

The ball travels a horizontal distance of 2·9 m to reach the top of the basket.

The effects of air resistance can be ignored.

(a) Calculate:

(i) the horizontal component of the initial velocity of the ball; **2**

(ii) the vertical component of the initial velocity of the ball. **2**

(b) Show that the time taken for the ball to reach the basket is 0·69 s. **2**

(c) Calculate the height h of the top of the basket. **3**

(d) A student observing the player makes the following statement.

"The player should throw the ball with a higher speed at the same angle. The ball would then land in the basket as before but it would take a shorter time to travel the 2·9 metres."

Explain why the student's statement is incorrect. **2**

Total marks 11

MARKS DO NOT WRITE IN THIS MARGIN

4. A force sensor is used to investigate the impact of a ball as it bounces on a flat horizontal surface. The ball has a mass of 0·050 kg and is dropped vertically, from rest, through a height of 1·6 m as shown.

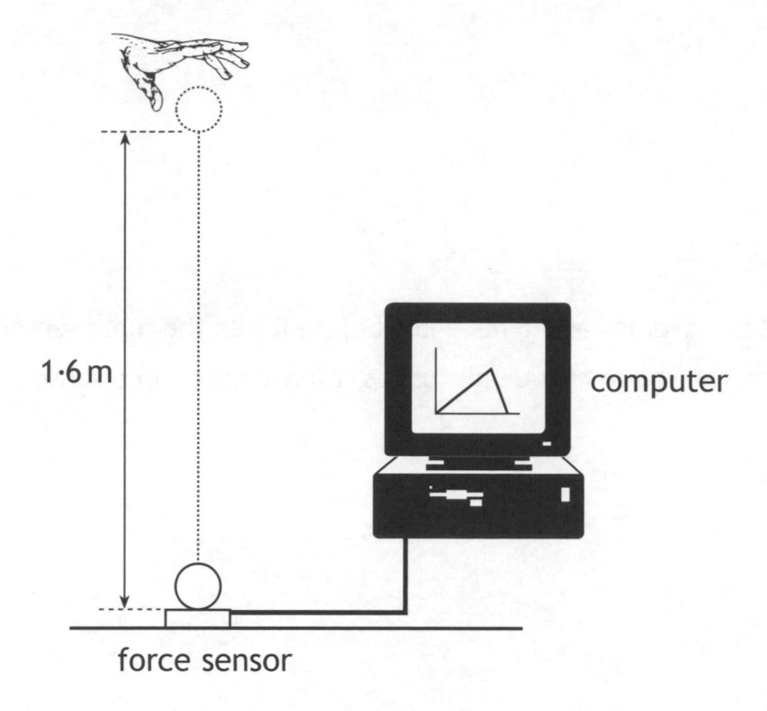

(a) The graph shows how the force on the ball varies with time during the impact.

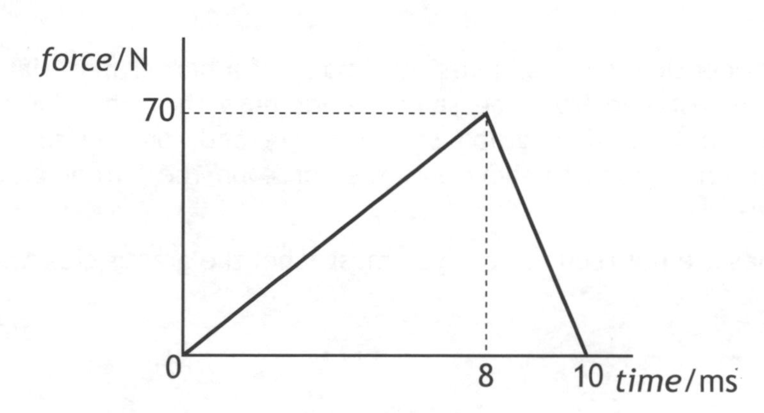

(i) Show by calculation that the magnitude of the impulse on the ball is 0·35 Ns. **2**

MARKS DO NOT WRITE IN THIS MARGIN

4. (a) (continued)

(ii) What is the magnitude and direction of the change in momentum of the ball? **2**

(iii) The ball is travelling at $5\cdot6\ ms^{-1}$ just before it hits the force sensor.

Calculate the speed of the ball just as it leaves the force sensor. **3**

(b) Another ball of identical size and mass, but made of a harder material, is dropped from rest and from the same height onto the same force sensor. Sketch the force-time graph shown above and, on the same axes, sketch another graph to show how the force on the harder ball varies with time. **2**

Numerical values are not required but you must label the graphs clearly.

Total marks 9

MARKS DO NOT WRITE IN THIS MARGIN

5. A train is stationary and a passenger on board hears a siren on another train approaching along a parallel track. The approaching train is travelling at a constant speed of $28{\cdot}0\ ms^{-1}$ and the siren produces a sound of frequency 294 Hz.

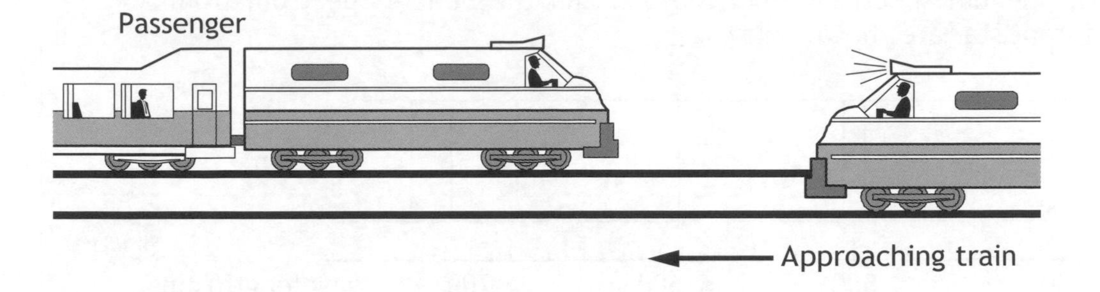

(a) (i) Explain, in terms of wavefronts, why the sound heard by the passenger does not have a frequency of 294 Hz. **2**

You may wish to include a diagram to support your answer.

(ii) Calculate the frequency of the sound heard by the passenger once the train has passed the passenger. **3**

MARKS | DO NOT WRITE IN THIS MARGIN

5. (continued)

(b) The spectrum of light from most stars contains lines corresponding to helium gas.

The helium spectrum from the Sun and the helium spectrum from a distant star are shown below.

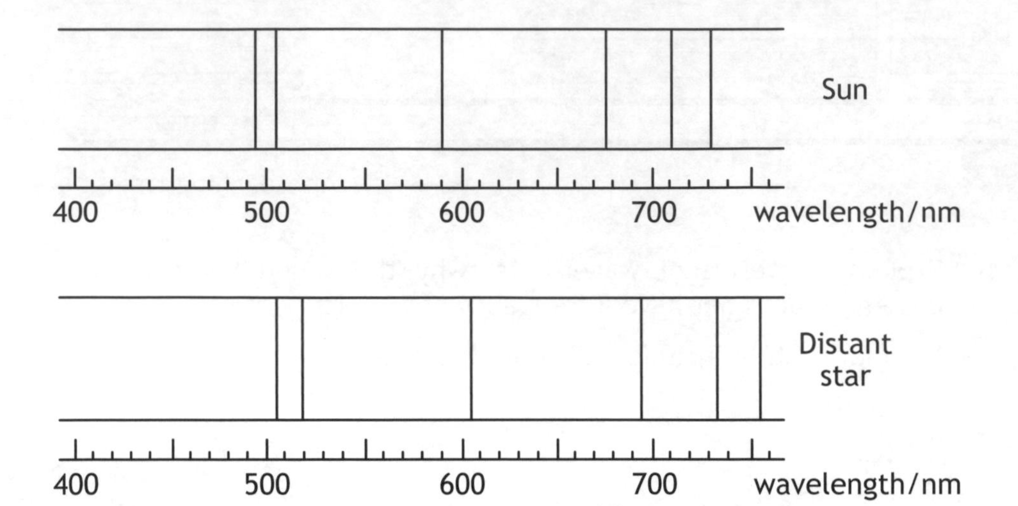

By comparing these spectra, what conclusion can be made about the distant star? Justify your answer. **2**

Total marks 7

MARKS DO NOT WRITE IN THIS MARGIN

6.

Two balls, one of mass 2kg and the other of mass 1kg, are dropped from the same height.

Use your knowledge of Physics to compare the time taken by each ball to reach the ground. **3**

MARKS DO NOT WRITE IN THIS MARGIN

7. (a) A point charge of +4·0 μC is shown.

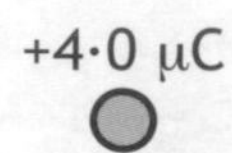

(i) Draw the electric field lines around the point charge. **1**

(ii) A point charge of −2·0 μC is now placed at a distance of 0·10 m from the first charge.

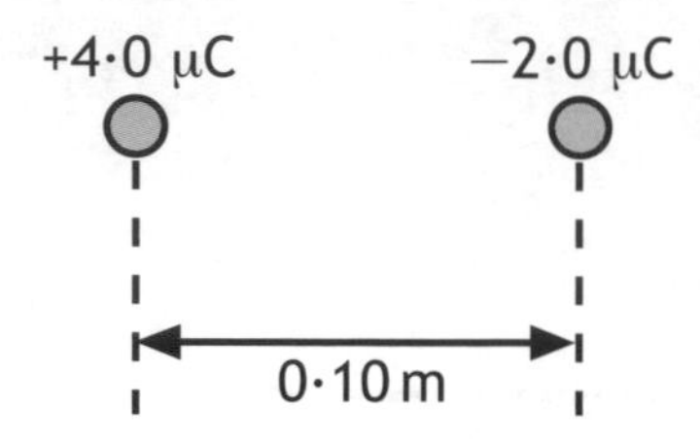

Explain why the electric field strength **is not** zero at any point between these two charges. **2**

(b) Two like charges experience a repulsive electrostatic force.

Explain why two protons in a nucleus do not fly apart. **2**

MARKS DO NOT WRITE IN THIS MARGIN

7. (continued)

(c) Information on the properties of three elementary particles together with the types of quarks and their corresponding antiquarks is shown below.

Properties of elementary particles			
Particle	Number of quarks	Charge	Baryon number
Proton	3	+e	1
Antiproton	3	−e	−1
Pi-meson	2	−e	0

Properties of quarks and antiquarks		
Particle	Charge	Baryon number
Up quark	$+\frac{2}{3}e$	$+\frac{1}{3}$
Down quark	$-\frac{1}{3}e$	$+\frac{1}{3}$
Anti-up quark	$-\frac{2}{3}e$	$-\frac{1}{3}$
Anti-down quark	$+\frac{1}{3}e$	$-\frac{1}{3}$

(i) Using information given, show that a proton consists of two up quarks and one down quark. **1**

(ii) State the combination of quarks that forms a pi-meson. **1**

Total marks 7

MARKS DO NOT WRITE IN THIS MARGIN

8. The diagram below shows the basic features of a proton accelerator. It is enclosed in an evacuated container.

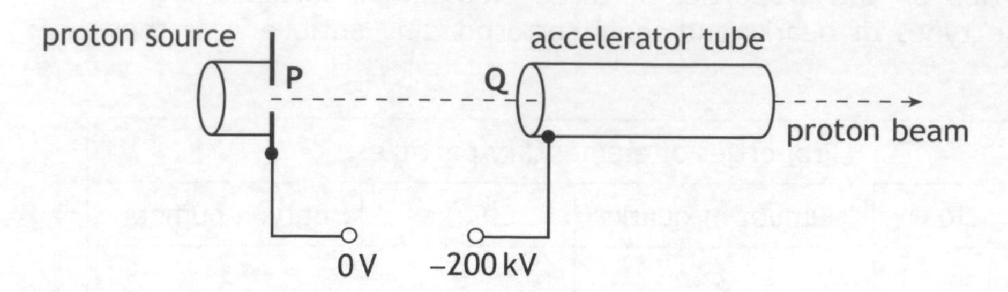

Protons released from the proton source start from rest at **P**.
A potential difference of 200 kV is maintained between **P** and **Q**.

(a) What is meant by the term *potential difference of 200 kV*? **1**

(b) **Explain** why protons released at **P** are accelerated towards **Q**. **1**

(c) **Calculate**:

(i) the work done on a proton as it accelerates from **P** to **Q**; **3**

(ii) the speed of a proton as it reaches **Q**. **3**

MARKS | DO NOT WRITE IN THIS MARGIN

8. (continued)

(d) The distance between **P** and **Q** is now halved.

What effect, if any, does this change have on the speed of a proton as it reaches **Q**? Justify your answer. **2**

Total marks 10

9. An experiment to determine the wavelength of light from a laser is shown.

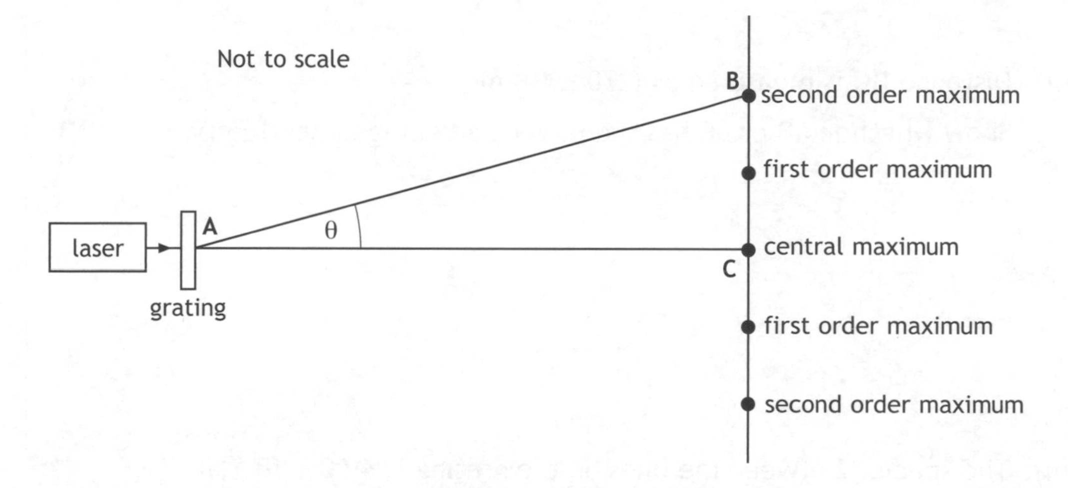

A **second** order maximum is observed at point **B**.

(a) **Explain** in terms of waves how a maximum is formed. **1**

MARKS DO NOT WRITE IN THIS MARGIN

9. (continued)

(b) Distance **AB** is measured six times.

The results are shown.

1·11 m   1·08 m   1·10 m   1·13 m   1·11 m   1·07 m

(i) **Calculate:**

(A) the mean value for distance **AB**; **2**

(B) the approximate random uncertainty in this value. **2**

(ii) Distance **BC** is measured as (270 ± 10) mm.

Show whether **AB** or **BC** has the larger percentage uncertainty. **3**

(iii) The spacing between the lines on the grating is $4{\cdot}00 \times 10^{-6}$ m.

Calculate the wavelength of the light from the laser.

Express your answer in the form

wavelength ± **absolute** uncertainty **4**

Total marks 12

MARKS DO NOT WRITE IN THIS MARGIN

10. A metal plate emits electrons when certain wavelengths of electromagnetic radiation are incident on it.

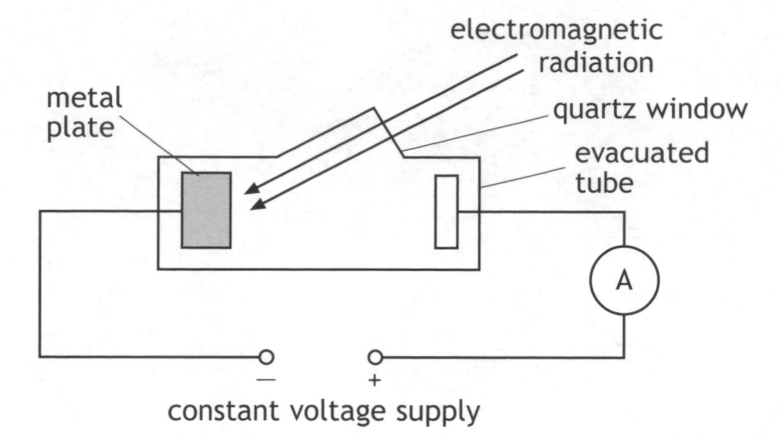

When light of wavelength 605 nm is incident on the metal plate, electrons are released with zero kinetic energy.

(a) Show that the work function of this metal is $3 \cdot 29 \times 10^{-19}$ J. **3**

(b) The wavelength of the incident radiation is now altered. Photons of energy $5 \cdot 12 \times 10^{-19}$ J are incident on the metal plate.

(i) Calculate the maximum kinetic energy of the electrons just as they leave the metal plate. **2**

(ii) The irradiance of this radiation on the metal plate is now decreased.

State the effect this has on the ammeter reading.

Justify your answer. **2**

Total marks 7

MARKS | DO NOT WRITE IN THIS MARGIN

11. A ray of red light is incident on a semicircular block of glass at the mid point of XY as shown.

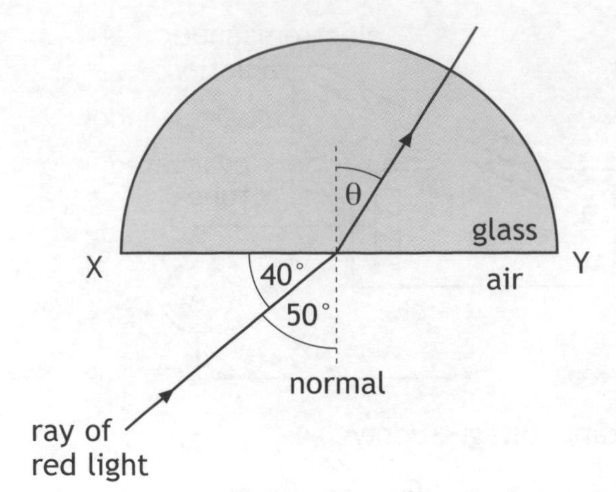

The refractive index of the block is 1·50 for this red light.

(a) Calculate angle θ shown on the diagram. **3**

(b) The wavelength of the red light **in the glass** is 420 nm. **3**

Calculate the wavelength of the light in air.

(c) The ray of red light is replaced by a ray of blue light incident at the same angle. The blue light enters the block at the same point.

Explain why the path taken by the blue light in the block is different to that taken by the red light. **1**

Total marks 7

MARKS DO NOT WRITE IN THIS MARGIN

12. An uncharged 2200 μF capacitor is connected in a circuit as shown.

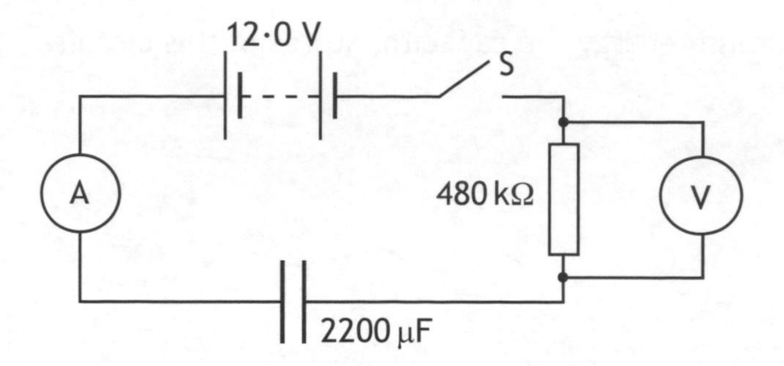

The battery has negligible internal resistance.

(a) Switch S is closed. Calculate the initial charging current. **3**

(b) At one instant during the charging process the potential difference **across the resistor** is 3·8 V.

Calculate the charge stored in the capacitor at this instant. **4**

MARKS DO NOT WRITE IN THIS MARGIN

12. (continued)

(c) Calculate the **maximum** energy the capacitor stores in this circuit. **3**

Total marks 10

13. A microphone is connected to the input terminals of an oscilloscope.

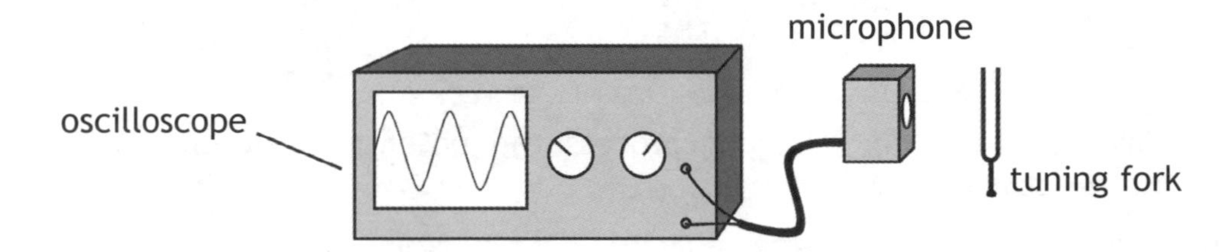

A tuning fork is made to vibrate and held close to the microphone as shown. The following diagram shows the trace obtained and the settings on the oscilloscope.

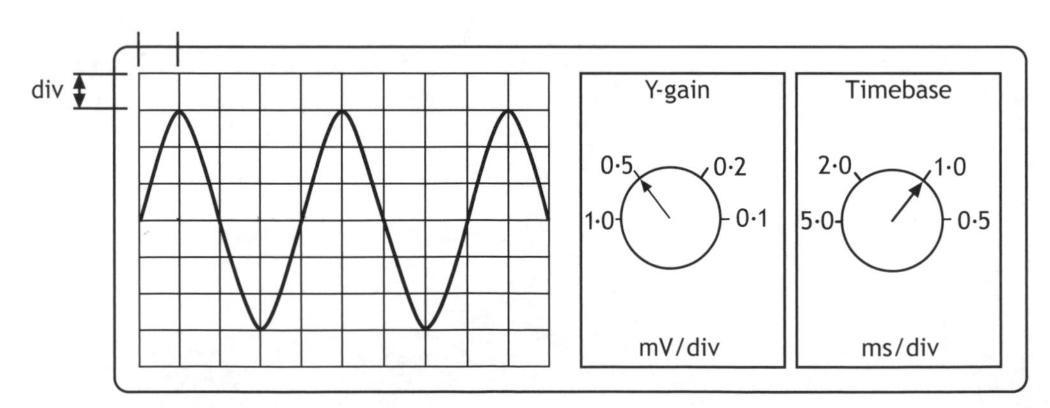

Calculate:

(a) the peak voltage of the signal; **1**

MARKS | DO NOT WRITE IN THIS MARGIN

13. (continued)

(b) the frequency of the signal. **3**

Total marks 4

14. (a) The circuit below was set up and the switch moved continually between A to B.

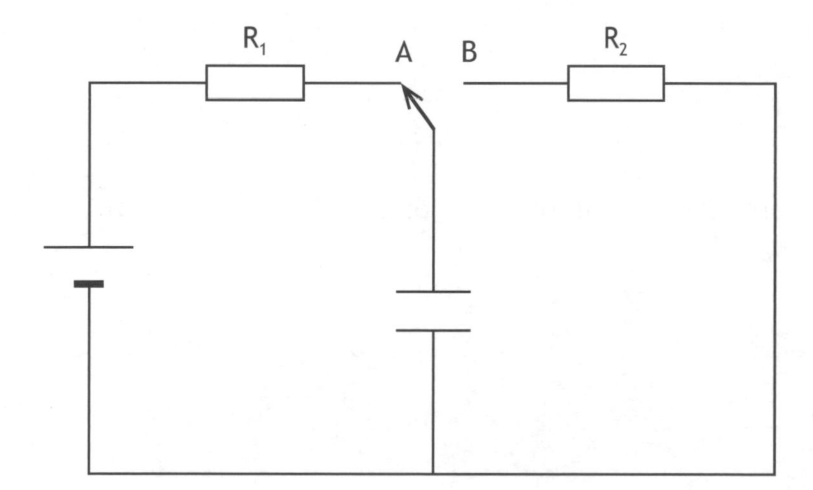

The graph below plots the current against time during the process.

MARKS DO NOT WRITE IN THIS MARGIN

14. (a) (continued)

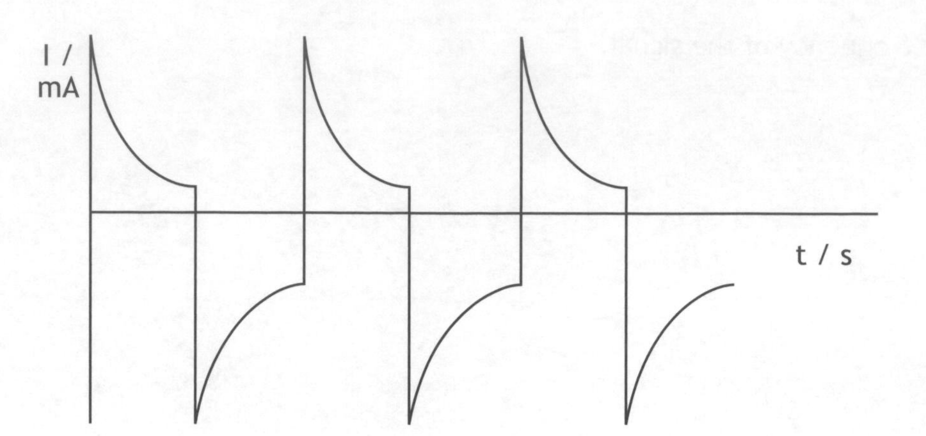

Use your knowledge of Physics to explain the shape of the graph. **3**

(b) Sketch the graph obtained if the rate of switching between A and B increased. **2**

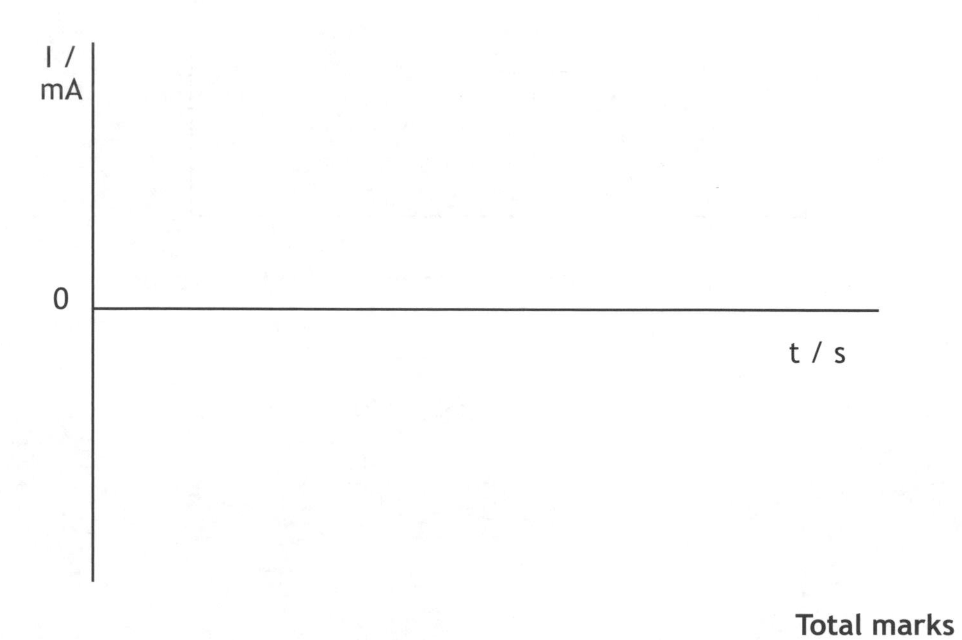

Total marks 5

[END OF MODEL PAPER]

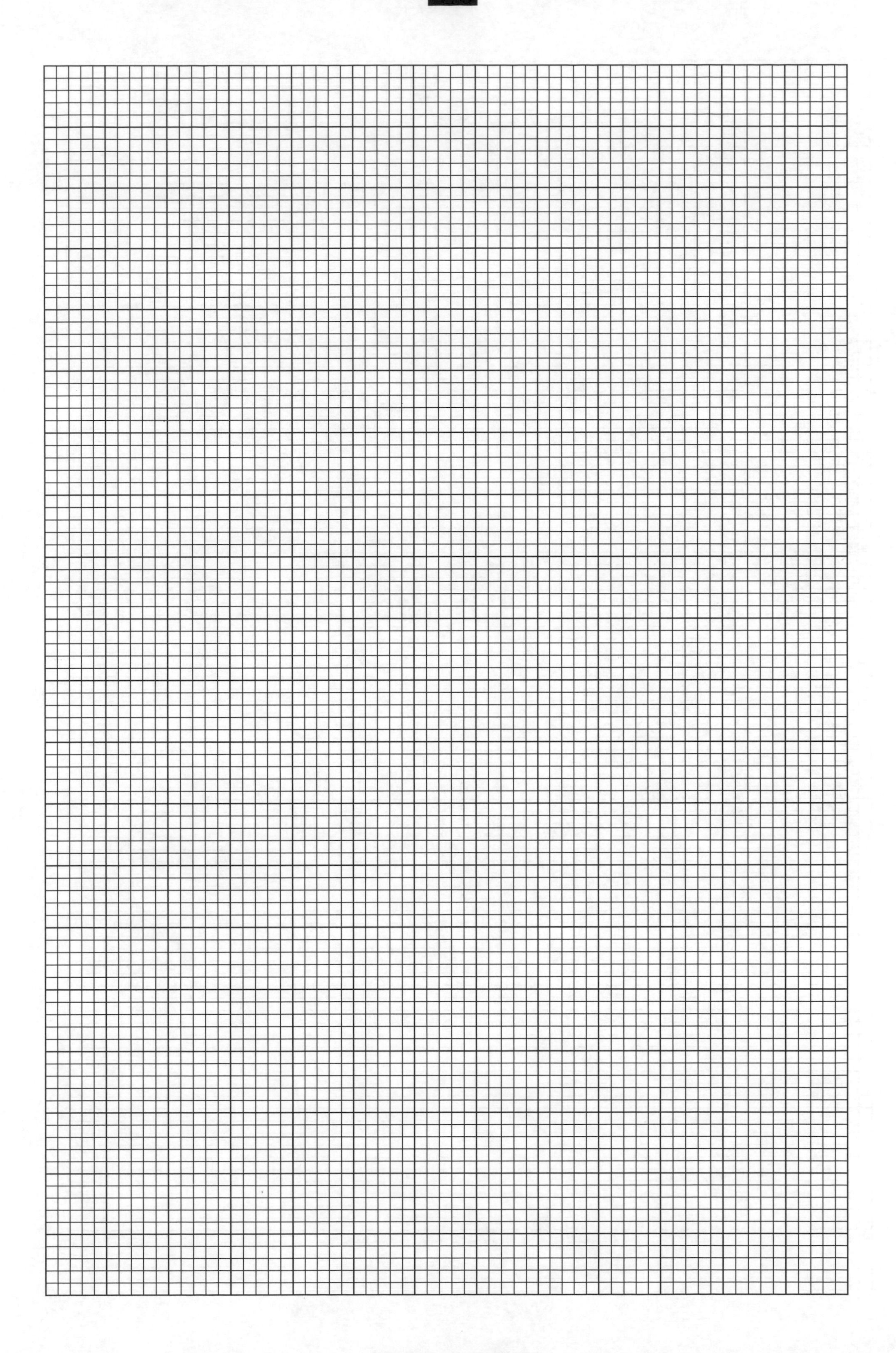

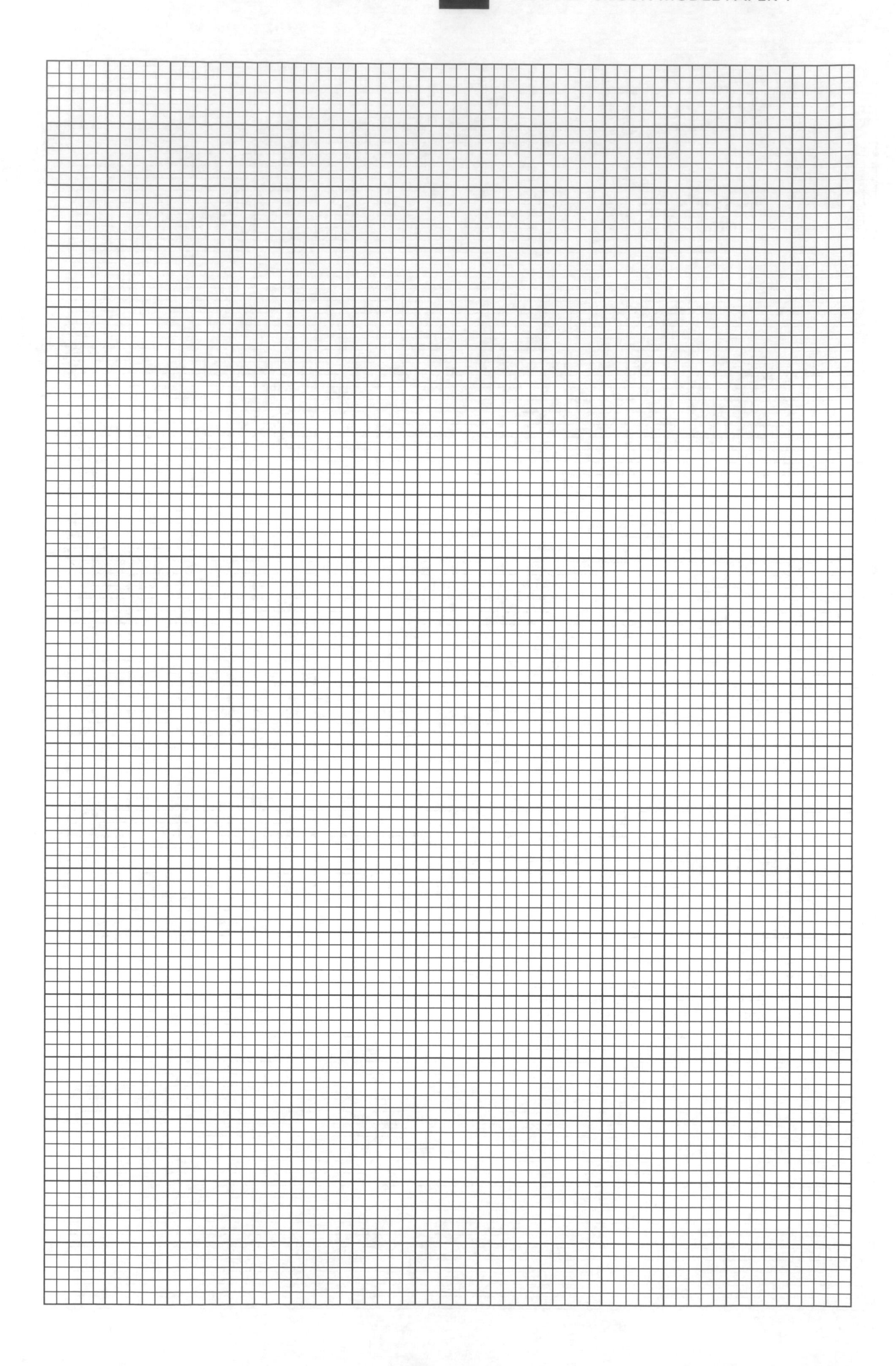

ADDITIONAL SPACE FOR ANSWERS AND ROUGH WORK

MARKS | DO NOT WRITE IN THIS MARGIN

ADDITIONAL SPACE FOR ANSWERS AND ROUGH WORK

MARKS | DO NOT WRITE IN THIS MARGIN

HIGHER FOR CfE

Model Paper 2

Whilst this Model Paper has been specially commissioned by Hodder Gibson for use as practice for the Higher (for Curriculum for Excellence) exams, the key reference documents remain the SQA Specimen Paper 2014 and SQA Past Paper 2015.

Physics
Section 1—Questions

Duration — 2 hours and 30 minutes

Instructions for the completion of Section 1 are given on *Page two* of your question and answer booklet.

Record your answers on the answer grid on *Page three* of your question and answer booklet.

Reference may be made to the Data Sheet on *Page two* of this booklet and to the Relationships Sheet.

Before leaving the examination room you must give your question and answer booklet to the Invigilator; if you do not, you may lose all the marks for this paper.

DATA SHEET

COMMON PHYSICAL QUANTITIES

Quantity	*Symbol*	*Value*	*Quantity*	*Symbol*	*Value*
Speed of light in vacuum	c	$3{\cdot}00 \times 10^{8}\,\mathrm{m\,s^{-1}}$	Planck's constant	h	$6{\cdot}63 \times 10^{-34}\,\mathrm{J\,s}$
Magnitude of the charge on an electron	e	$1{\cdot}60 \times 10^{-19}\,\mathrm{C}$	Mass of electron	m_e	$9{\cdot}11 \times 10^{-31}\,\mathrm{kg}$
Universal Constant of Gravitation	G	$6{\cdot}67 \times 10^{-11}\,\mathrm{m^3\,kg^{-1}\,s^{-2}}$	Mass of neutron	m_n	$1{\cdot}675 \times 10^{-27}\,\mathrm{kg}$
Gravitational acceleration on Earth	g	$9{\cdot}8\,\mathrm{m\,s^{-2}}$	Mass of proton	m_p	$1{\cdot}673 \times 10^{-27}\,\mathrm{kg}$
Hubble's constant	H_0	$2{\cdot}3 \times 10^{-18}\,\mathrm{s^{-1}}$			

REFRACTIVE INDICES

The refractive indices refer to sodium light of wavelength 589 nm and to substances at a temperature of 273 K.

Substance	*Refractive index*	*Substance*	*Refractive index*
Diamond	2·42	Water	1·33
Crown glass	1·50	Air	1·00

SPECTRAL LINES

Element	*Wavelength/nm*	*Colour*
Hydrogen	656	Red
	486	Blue-green
	434	Blue-violet
	410	Violet
	397	Ultraviolet
	389	Ultraviolet
Sodium	589	Yellow

Element	*Wavelength/nm*	*Colour*
Cadmium	644	Red
	509	Green
	480	Blue

Lasers		
Element	*Wavelength/nm*	*Colour*
Carbon dioxide	9550 } 10590 }	Infrared
Helium-neon	633	Red

PROPERTIES OF SELECTED MATERIALS

Substance	*Density/kg m^{-3}*	*Melting Point/K*	*Boiling Point/K*
Aluminium	$2{\cdot}70 \times 10^{3}$	933	2623
Copper	$8{\cdot}96 \times 10^{3}$	1357	2853
Ice	$9{\cdot}20 \times 10^{2}$	273	....
Sea Water	$1{\cdot}02 \times 10^{3}$	264	377
Water	$1{\cdot}00 \times 10^{3}$	273	373
Air	1·29	....	....
Hydrogen	$9{\cdot}0 \times 10^{-2}$	14	20

The gas densities refer to a temperature of 273 K and a pressure of $1{\cdot}01 \times 10^{5}$ Pa.

SECTION 1 — 20 marks

Attempt ALL questions

1. A student makes five separate measurements of the diameter of a lens.

These measurements are shown in the table.

Diameter of lens/mm	22·5	22·6	22·4	22·6	22·9

The approximate random uncertainty in the mean value of the diameter is

A 0.1 mm

B 0.2 mm

C 0.3 mm

D 0.4 mm

E 0.5 mm.

2. A golfer strikes a golf ball which then moves off at an angle to the ground. The ball follows the path shown.

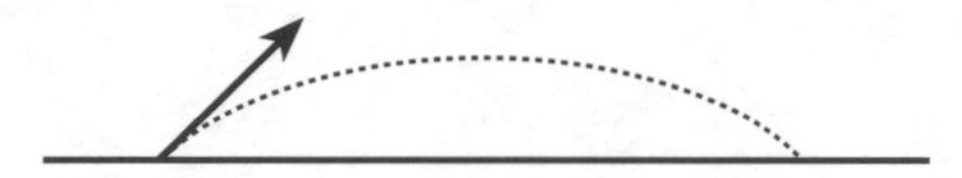

The graphs below show how the horizontal and vertical components of the velocity of the ball vary with time.

What is the speed of the ball just before it hits the ground?

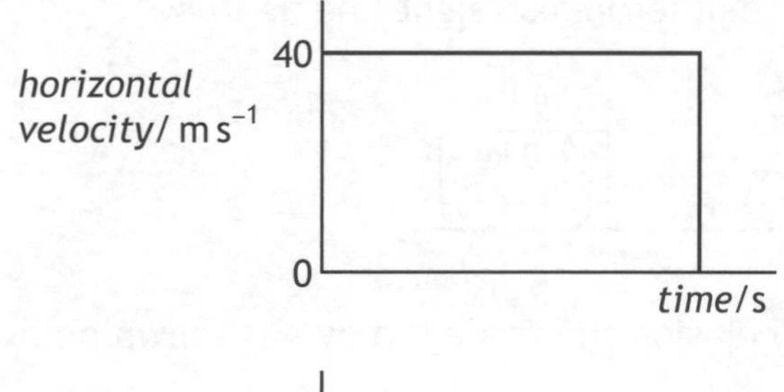

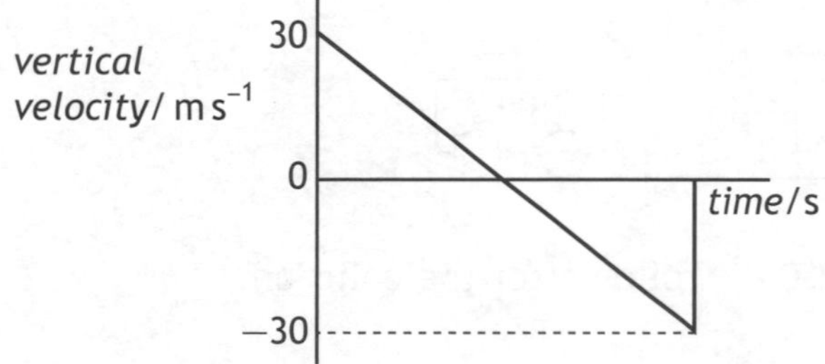

A 10 m s^{-1}

B 30 m s^{-1}

C 40 m s^{-1}

D 50 m s^{-1}

E 70 m s^{-1}

3. An object starts from rest and accelerates in a straight line.

The graph shows how the acceleration of the object varies with time.

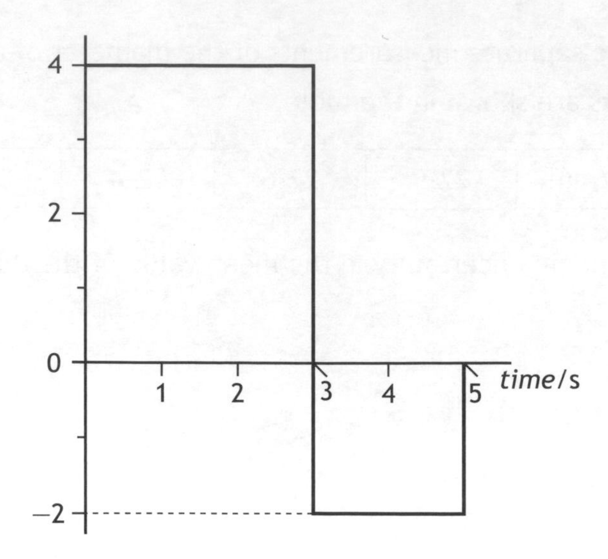

The object's speed at 5 seconds is

A $2\ \text{m s}^{-1}$

B $8\ \text{m s}^{-1}$

C $12\ \text{m s}^{-1}$

D $16\ \text{m s}^{-1}$

E $20\ \text{m s}^{-1}$.

4. Two trolleys travel towards each other in a straight line as shown.

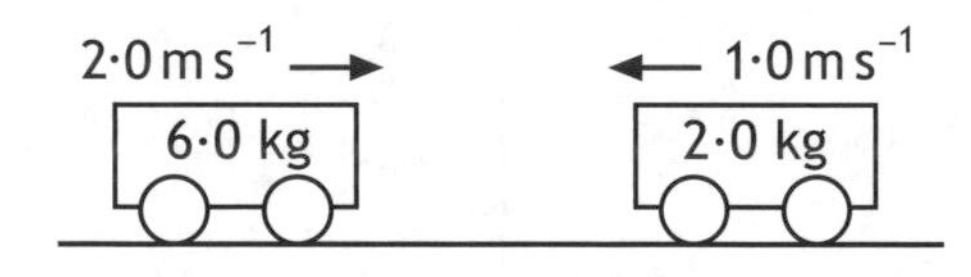

The trolleys collide. After the collision the trolleys move as shown below.

1·0 m s⁻¹ → 6·0 kg

v → 2·0 kg

What is the speed v of the 2.0 kg trolley after the collision?

A $1.25\ \text{m s}^{-1}$

B $1.75\ \text{m s}^{-1}$

C $2.0\ \text{m s}^{-1}$

D $4.0\ \text{m s}^{-1}$

E $5.0\ \text{m s}^{-1}$

5. An object appears to have its length halved when viewed from a stationary frame of reference. It must be travelling at

A 0·99c

B 0·48c

C 0·76c

D 0·91c

E 0·87c

6. A spectral line from light from a binary star is observed to have a wavelength of 450 nm. The same line on earth has a wavelength of 472 nm.

At that point the star is:

A Moving away from earth at $1·5 \times 10^7$ m s^{-1}

B Moving towards earth $1·5 \times 10^7$ m s^{-1}

C Moving towards earth $1·5 \times 10^8$ m s^{-1}

D Moving away from earth at $1·5 \times 10^7$ m s^{-1}

E Moving towards earth at $2·1 \times 10^8$ m s^{-1}.

7. An ambulance emits a note of frequency of 460 Hz when stationary.

The ambulance, travelling at a constant speed of 20 m s^{-1}, passes an observer.

The frequency heard by the observer as the **ambulance approaches** is

A 424 Hz

B 452 Hz

C 489 Hz

D 470Hz

E 434 Hz.

8. Which row of the table gives the correct exchange particles for the fundamental forces?

	Electromagnetic	Weak	Strong
A	W, Z boson	Photon	Gluon
B	W, Z boson	Gluon	Photon
C	Photon	Gluon	W, Z boson
D	Gluon	Photon	W, Z boson
E	Photon	W, Z boson	Gluon

9. The product, X, of a nuclear reaction passes through an electric field as shown.

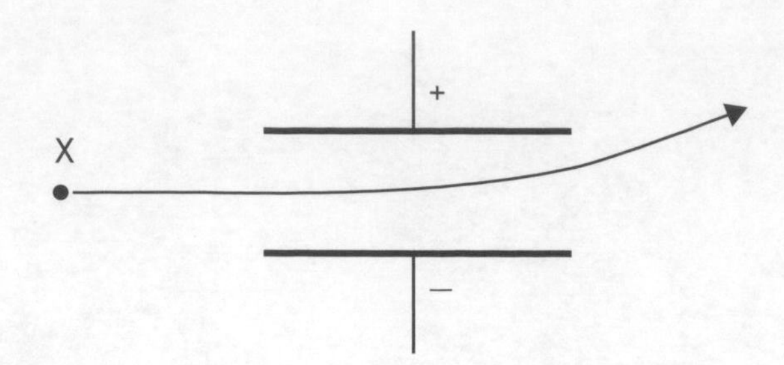

Product X is

A an alpha particle

B a beta particle

C gamma radiation

D a fast neutron

E a slow neutron.

10. The following statement describes a fusion reaction.

$$^{2}_{1}H + ^{2}_{1}H \longrightarrow ^{3}_{2}He + ^{1}_{0}n + \text{energy}$$

The total mass of the particles before the reaction is $6{\cdot}684 \times 10^{-27}$ kg.

The total mass of the particles after the reaction is $6{\cdot}680 \times 10^{-27}$ kg.

The energy released in this reaction is

A $6{\cdot}012 \times 10^{-10}$ J

B $6{\cdot}016 \times 10^{-10}$ J

C $1{\cdot}800 \times 10^{-13}$ J

D $3{\cdot}600 \times 10^{-13}$ J

E $1{\cdot}200 \times 10^{-21}$ J.

11. Ultraviolet radiation causes the emission of photoelectrons from a zinc plate.

The Irradiance of the ultraviolet radiation is increased. Which row in the following table shows the effect of this change?

	Maximum kinetic energy of a photoelectron	*Number of photoelectrons per second*
A	increases	no change
B	no change	increases
C	no change	no change
D	increases	increases
E	decreases	increases

12. A source of microwaves of wavelength λ is placed behind two slits, R and S.

A microwave detector records a maximum when it is placed at P.

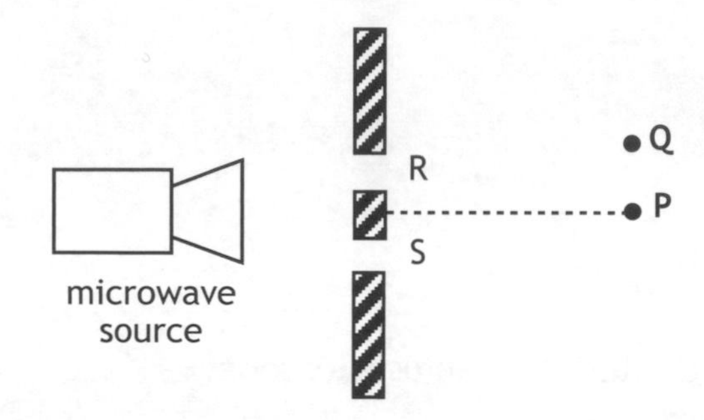

The detector is moved and the next maximum is recorded at Q.

The path difference (SQ − RQ) is

A 0

B $\frac{\lambda}{2}$

C λ

B $\frac{3\lambda}{2}$

E 2λ.

13. The value of the absolute refractive index of diamond is 2·42.

The critical angle for diamond is

A 0·413°

B 24·4°

C 42·0°

D 65·6°

E 90·0°

14. The diagram represents some electron transitions between energy levels in an atom.

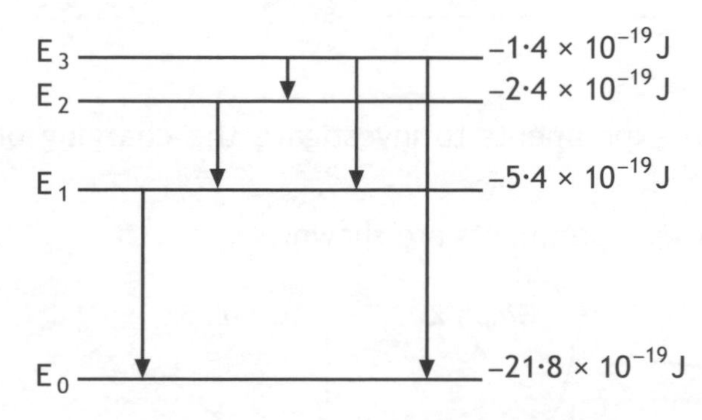

The radiation emitted with the shortest wavelength is produced by an electron making transition

14. (continued)

A E_1 to E_0

B E_2 to E_1

C E_3 to E_2

D E_3 to E_1

E E_3 to E_0.

15. A signal from a power supply is displayed on an oscilloscope.

The trace on the oscilloscope is shown.

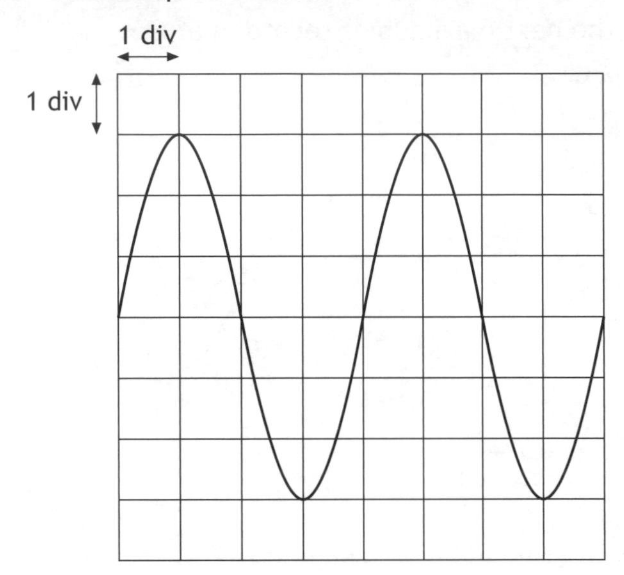

The time-base is set at 0.01 s/div and the Y-gain is set at 4.0 V/div.

Which row in the table shows the r.m.s. voltage and the frequency of the signal?

	r.m.s voltage/V	*frequency*/Hz
A	8·5	25
B	12	25
C	24	25
D	8·5	50
E	12	50

16. A student carries out three experiments to investigate the charging of a capacitor using a d.c. supply.

The graphs obtained from the experiments are shown.

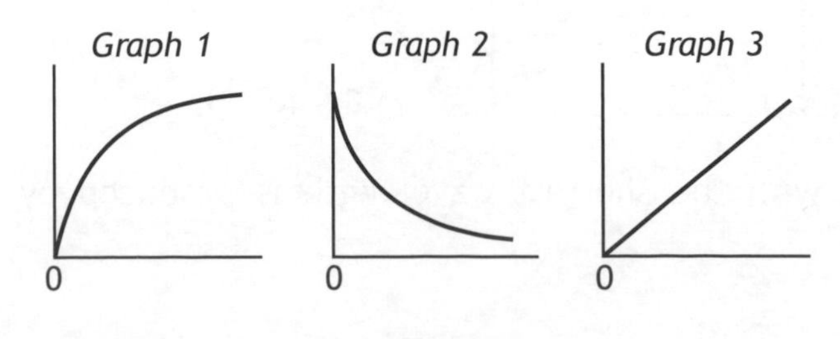

16. (continued)

The axes of the graphs have not been labelled.

Which row in the table shows the labels for the axes of the graphs?

	Graph 1	*Graph 2*	*Graph 3*
A	voltage and time	current and time	charge and voltage
B	current and time	voltage and time	charge and voltage
C	current and time	charge and voltage	voltage and time
D	charge and voltage	current and time	voltage and time
E	voltage and time	charge and voltage	current and time

17. Which of the following combinations of resistors has the greatest resistance between X and Y?

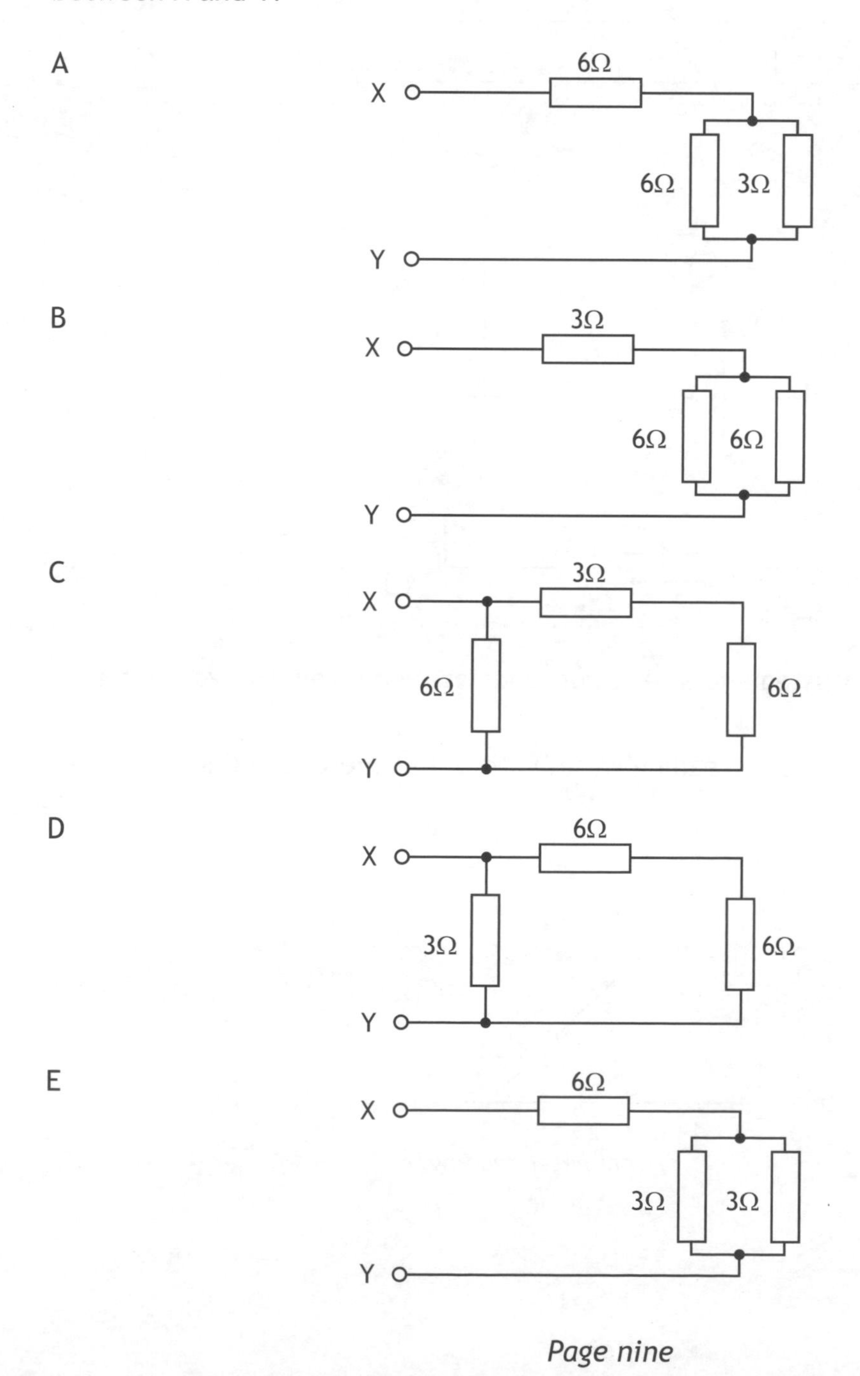

18. In an experiment to find the capacitance of a capacitor, a student makes the following measurements.

potential difference across capacitor = (10·0 ± 0·1) V

charge stored by capacitor = (500 ± 25) μC

Which row in the table gives the capacitance of the capacitor and the percentage uncertainty in the capacitance?

	Capacitance/μF	*Percentage uncertainty*
A	0·02	1
B	0·02	5
C	50	1
D	50	5
E	5000	6

19. A circuit is set up as shown.

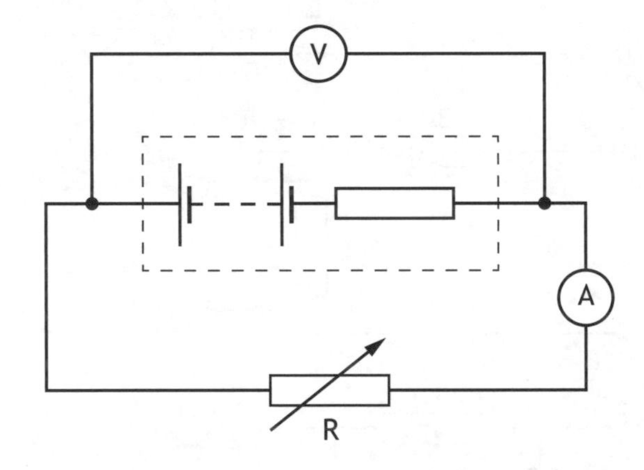

The variable resistor R is adjusted and a series of readings taken from the voltmeter and ammeter.

The graph shows how the voltmeter reading varies with the ammeter reading.

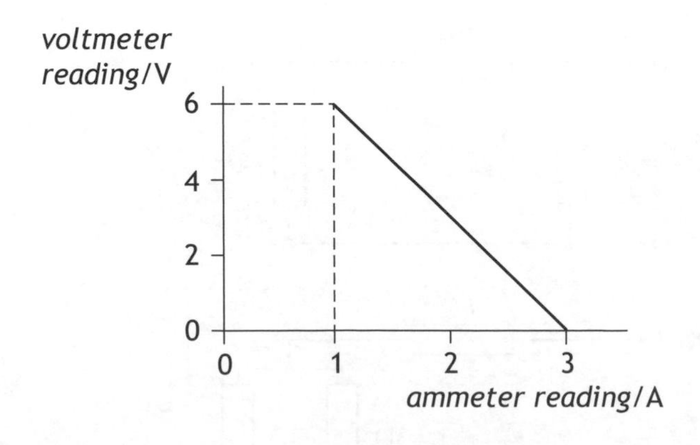

19. (continued)

Which row in the table shows the values for the e.m.f. and internal resistance of the battery in the circuit?

	e.m.f./V	*internal resistance/Ω*
A	6	2
B	6	3
C	9	2
D	9	3
E	9	6

20. In the following circuit, the supply has negligible internal resistance.

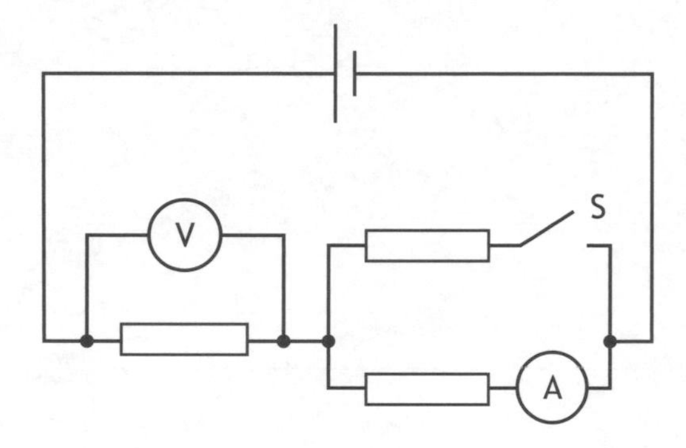

Switch S is now closed.

Which row in the table shows the effect on the ammeter and voltmeter readings?

	Ammenter reading	*Voltmeter reading*
A	increases	increases
B	increases	decreases
C	decreases	decreases
D	decreases	increases
E	decreases	remains the same

[END OF SECTION 1. NOW ATTEMPT THE QUESTIONS IN SECTION 2 OF YOUR QUESTION AND ANSWER BOOKLET]

National
Qualifications
MODEL PAPER 2

Physics
Relationships Sheet

Date — Not applicable

Relationships required for Physics Higher

$d = \bar{v}t$

$s = \bar{v}t$

$v = u + at$

$s = ut + \frac{1}{2}at^2$

$v^2 = u^2 + 2as$

$s = \frac{1}{2}(u + v)t$

$W = mg$

$F = ma$

$E_W = Fd$

$E_p = mgh$

$E_k = \frac{1}{2}mv^2$

$P = \frac{E}{t}$

$p = mv$

$Ft = mv - mu$

$F = G\frac{m_1 m_2}{r^2}$

$t' = \frac{t}{\sqrt{1 - \left(\frac{v}{c}\right)^2}}$

$l' = l\sqrt{1 - \left(\frac{v}{c}\right)^2}$

$f_o = f_s\left(\frac{v}{v \pm v_s}\right)$

$z = \frac{\lambda_{observed} - \lambda_{rest}}{\lambda_{rest}}$

$z = \frac{v}{c}$

$v = H_0 d$

$W = QV$

$E = mc^2$

$E = hf$

$E_k = hf - hf_0$

$E_2 - E_1 = hf$

$T = \frac{1}{f}$

$v = f\lambda$

$d\sin\theta = m\lambda$

$n = \frac{\sin\theta_1}{\sin\theta_2}$

$\frac{\sin\theta_1}{\sin\theta_2} = \frac{\lambda_1}{\lambda_2} = \frac{v_1}{v_2}$

$\sin\theta_c = \frac{1}{n}$

$I = \frac{k}{d^2}$

$I = \frac{P}{A}$

$V_{peak} = \sqrt{2}V_{rms}$

$I_{peak} = \sqrt{2}I_{rms}$

$Q = It$

$V = IR$

$P = IV = I^2R = \frac{V^2}{R}$

$R_T = R_1 + R_2 + \ldots$

$\frac{1}{R_T} = \frac{1}{R_1} + \frac{1}{R_2} + \ldots$

$E = V + Ir$

$V_1 = \left(\frac{R_1}{R_1 + R_2}\right)V_s$

$\frac{V_1}{V_2} = \frac{R_1}{R_2}$

$C = \frac{Q}{V}$

$E = \frac{1}{2}QV = \frac{1}{2}CV^2 = \frac{1}{2}\frac{Q^2}{C}$

path difference $= m\lambda$ or $\left(m + \frac{1}{2}\right)\lambda$ where $m = 0, 1, 2 \ldots$

random uncertainty $= \frac{\text{max. value} - \text{min. value}}{\text{number of values}}$

Additional Relationships

Circle

circumference = $2\pi r$

area = πr^2

Sphere

area = $4\pi r^2$

volume = $\frac{4}{3}\pi r^3$

Trigonometry

$$\sin\Theta = \frac{\text{opposite}}{\text{hypotenuse}}$$

$$\cos\Theta = \frac{\text{adjacent}}{\text{hypotenuse}}$$

$$\tan\Theta = \frac{\text{opposite}}{\text{adjacent}}$$

$$\sin^2\Theta + \cos^2\Theta = 1$$

Electron Arrangements of Elements

Key

Atomic number Symbol Electron arrangement Name

Transition Elements

Group 1 (1)	Group 2 (2)	(3)	(4)	(5)	(6)	(7)	(8)	(9)	(10)	(11)	(12)	Group 3 (13)	Group 4 (14)	Group 5 (15)	Group 6 (16)	Group 7 (17)	Group 0 (18)
1 **H** 1 Hydrogen																	2 **He** 2 Helium
3 **Li** 2,1 Lithium	4 **Be** 2,2 Beryllium											5 **B** 2,3 Boron	6 **C** 2,4 Carbon	7 **N** 2,5 Nitrogen	8 **O** 2,6 Oxygen	9 **F** 2,7 Fluorine	10 **Ne** 2,8 Neon
11 **Na** 2,8,1 Sodium	12 **Mg** 2,8,2 Magnesium											13 **Al** 2,8,3 Aluminium	14 **Si** 2,8,4 Silicon	15 **P** 2,8,5 Phosphorus	16 **S** 2,8,6 Sulfur	17 **Cl** 2,8,7 Chlorine	18 **Ar** 2,8,8 Argon
19 **K** 2,8,8,1 Potassium	20 **Ca** 2,8,8,2 Calcium	21 **Sc** 2,8,9,2 Scandium	22 **Ti** 2,8,10,2 Titanium	23 **V** 2,8,11,2 Vanadium	24 **Cr** 2,8,13,1 Chromium	25 **Mn** 2,8,13,2 Manganese	26 **Fe** 2,8,14,2 Iron	27 **Co** 2,8,15,2 Cobalt	28 **Ni** 2,8,16,2 Nickel	29 **Cu** 2,8,18,1 Copper	30 **Zn** 2,8,18,2 Zinc	31 **Ga** 2,8,18,3 Gallium	32 **Ge** 2,8,18,4 Germanium	33 **As** 2,8,18,5 Arsenic	34 **Se** 2,8,18,6 Selenium	35 **Br** 2,8,18,7 Bromine	36 **Kr** 2,8,18,8 Krypton
37 **Rb** 2,8,18,8,1 Rubidium	38 **Sr** 2,8,18,8,2 Strontium	39 **Y** 2,8,18,9,2 Yttrium	40 **Zr** 2,8,18, 10,2 Zirconium	41 **Nb** 2,8,18, 12,1 Niobium	42 **Mo** 2,8,18,13, 1 Molybdenum	43 **Tc** 2,8,18,13, 2 Technetium	44 **Ru** 2,8,18,15, 1 Ruthenium	45 **Rh** 2,8,18,16, 1 Rhodium	46 **Pd** 2,8,18, 18,0 Palladium	47 **Ag** 2,8,18, 18,1 Silver	48 **Cd** 2,8,18, 18,2 Cadmium	49 **In** 2,8,18, 18,3 Indium	50 **Sn** 2,8,18, 18,4 Tin	51 **Sb** 2,8,18, 18,5 Antimony	52 **Te** 2,8,18, 18,6 Tellurium	53 **I** 2,8,18, 18,7 Iodine	54 **Xe** 2,8,18, 18,8 Xenon
55 **Cs** 2,8,18,18, 8,1 Caesium	56 **Ba** 2,8,18,18, 8,2 Barium	57 **La** 2,8,18,18, 9,2 Lanthanum	72 **Hf** 2,8,18,32, 10,2 Hafnium	73 **Ta** 2,8,18, 32,11,2 Tantalum	74 **W** 2,8,18,32, 12,2 Tungsten	75 **Re** 2,8,18,32, 13,2 Rhenium	76 **Os** 2,8,18,32, 14,2 Osmium	77 **Ir** 2,8,18,32, 15,2 Iridium	78 **Pt** 2,8,18,32, 17,1 Platinum	79 **Au** 2,8,18, 32,18,1 Gold	80 **Hg** 2,8,18, 32,18,2 Mercury	81 **Tl** 2,8,18, 32,18,3 Thallium	82 **Pb** 2,8,18, 32,18,4 Lead	83 **Bi** 2,8,18, 32,18,5 Bismuth	84 **Po** 2,8,18, 32,18,6 Polonium	85 **At** 2,8,18, 32,18,7 Astatine	86 **Rn** 2,8,18, 32,18,8 Radon
87 **Fr** 2,8,18,32, 18,8,1 Francium	88 **Ra** 2,8,18,32, 18,8,2 Radium	89 **Ac** 2,8,18,32, 18,9,2 Actinium	104 **Rf** 2,8,18,32, 32,10,2 Rutherfordium	105 **Db** 2,8,18,32, 32,11,2 Dubnium	106 **Sg** 2,8,18,32, 32,12,2 Seaborgium	107 **Bh** 2,8,18,32, 32,13,2 Bohrium	108 **Hs** 2,8,18,32, 32,14,2 Hassium	109 **Mt** 2,8,18,32, 32,15,2 Meitnerium	110 **Ds** 2,8,18,32, 32,17,1 Darmstadtium	111 **Rg** 2,8,18,32, 32,18,1 Roentgenium	112 **Cn** 2,8,18,32, 32,18,2 Copernicium						

Lanthanides	57 **La** 2,8,18, 18,9,2 Lanthanum	58 **Ce** 2,8,18, 20,8,2 Cerium	59 **Pr** 2,8,18,21, 8,2 Praseodymium	60 **Nd** 2,8,18,22, 8,2 Neodymium	61 **Pm** 2,8,18,23, 8,2 Promethium	62 **Sm** 2,8,18,24, 8,2 Samarium	63 **Eu** 2,8,18,25, 8,2 Europium	64 **Gd** 2,8,18,25, 9,2 Gadolinium	65 **Tb** 2,8,18,27, 8,2 Terbium	66 **Dy** 2,8,18,28, 8,2 Dysprosium	67 **Ho** 2,8,18,29, 8,2 Holmium	68 **Er** 2,8,18,30, 8,2 Erbium	69 **Tm** 2,8,18,31, 8,2 Thulium	70 **Yb** 2,8,18,32, 8,2 Ytterbium	71 **Lu** 2,8,18,32, 9,2 Lutetium
Actinides	89 **Ac** 2,8,18,32, 18,9,2 Actinium	90 **Th** 2,8,18,32, 18,10,2 Thorium	91 **Pa** 2,8,18,32, 20,9,2 Protactinium	92 **U** 2,8,18,32, 21,9,2 Uranium	93 **Np** 2,8,18,32, 22,9,2 Neptunium	94 **Pu** 2,8,18,32, 24,8,2 Plutonium	95 **Am** 2,8,18,32, 25,8,2 Americium	96 **Cm** 2,8,18,32, 25,9,2 Curium	97 **Bk** 2,8,18,32, 27,8,2 Berkelium	98 **Cf** 2,8,18,32, 28,8,2 Californium	99 **Es** 2,8,18,32, 29,8,2 Einsteinium	100 **Fm** 2,8,18,32, 30,8,2 Fermium	101 **Md** 2,8,18,32, 31,8,2 Mendelevium	102 **No** 2,8,18,32, 32,8,2 Nobelium	103 **Lr** 2,8,18,32, 32,9,2 Lawrencium

Physics
Section 1 — Answer Grid and Section 2

Duration — 2 hours and 30 minutes

Fill in these boxes and read what is printed below.

Full name of centre

Town

Forename(s)

Surname

Number of seat

Date of birth

Day Month Year

D D M M Y Y

Scottish candidate number

Total marks — 130

SECTION 1 — 20 marks

Attempt ALL questions.

Instructions for the completion of Section 1 are given on *Page two*.

SECTION 2 — 110 marks

Attempt ALL questions.

Reference may be made to the Data Sheet on *Page two* of the question paper and to the Relationship Sheet.

Write your answers clearly in the spaces provided in this booklet. Additional space for answers and rough work is provided at the end of this booklet. If you use this space you must clearly identify the question number you are attempting. Any rough work must be written in this booklet. You should score through your rough work when you have written your final copy.

Use **blue** or **black** ink.

Care should be taken to give an appropriate number of significant figures in the final answers to calculations.

Before leaving the examination room you must give this booklet to the Invigilator; if you do not you may lose all the marks for this paper.

HODDER GIBSON
LEARN MORE

SECTION 1 — 20 marks

The questions for Section 1 are contained in the booklet Physics Section 1 – Questions.
Read these and record your answers on the answer grid on *Page three* opposite.
Do **NOT** use gel pens.

1. The answer to each question is **either** A, B, C, D or E. Decide what your answer is, then fill in the appropriate bubble (see sample question below).

2. There is **only one correct** answer to each question.

3. Any rough working should be done on the additional space for answers and rough work at the end of this booklet.

Sample Question

The energy unit measured by the electricity meter in your home is the:

A ampere
B kilowatt-hour
C watt
D coulomb
E volt.

The correct answer is **B**—kilowatt-hour. The answer **B** bubble has been clearly filled in (see below).

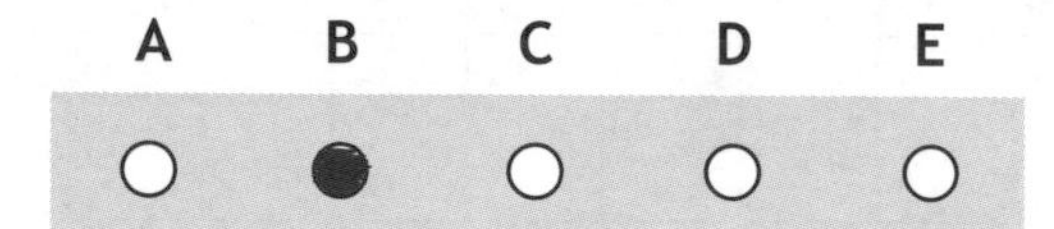

Changing an answer

If you decide to change your answer, cancel your first answer by putting a cross through it (see below) and fill in the answer you want. The answer below has been changed to **D**.

A B C D E

If you then decide to change back to an answer you have already scored out, put a tick (✓) to the **right** of the answer you want, as shown below:

A B C D E ✓

or

A B C D E ✓

SECTION 1 — Answer Grid

	A	B	C	D	E
1	O	O	O	O	O
2	O	O	O	O	O
3	O	O	O	O	O
4	O	O	O	O	O
5	O	O	O	O	O
6	O	O	O	O	O
7	O	O	O	O	O
8	O	O	O	O	O
9	O	O	O	O	O
10	O	O	O	O	O
11	O	O	O	O	O
12	O	O	O	O	O
13	O	O	O	O	O
14	O	O	O	O	O
15	O	O	O	O	O
16	O	O	O	O	O
17	O	O	O	O	O
18	O	O	O	O	O
19	O	O	O	O	O
20	O	O	O	O	O

MARKS DO NOT WRITE IN THIS MARGIN

SECTION 2 — 110 marks

Attempt ALL questions

1. (a) A satellite of mass 295 kg orbits the earth at a distance of 245 km from the earth's centre

Mass of earth = $6{\cdot}0 \times 10^{24}$ kg

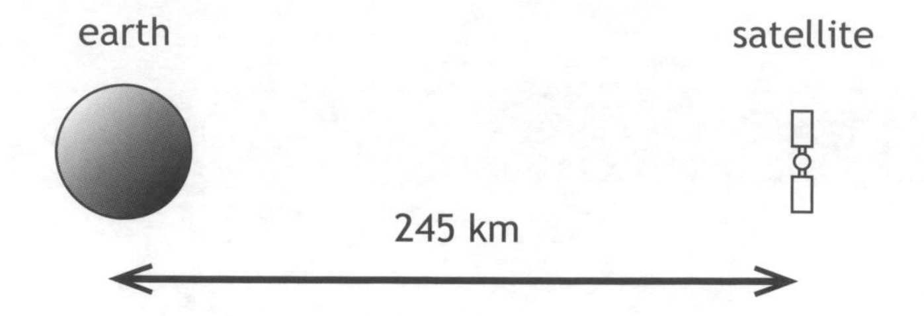

Calculate the force that keeps the satellite in orbit around the earth. **3**

(b) It can be shown that the relationship between the period T, in seconds of a satellite is given by

$$T = 2\pi\sqrt{\frac{r^3}{GM}}$$

where r = radius of the orbit

M = mass of the earth.

Calculate the height of a geostationary satellite above the earth. **4**

The period of a geostationary satellite is 24 hours.

Radius of the earth = $6{\cdot}4 \times 10^6$ m.

MARKS DO NOT WRITE IN THIS MARGIN

1. (continued)

(c) Cosmologists have come up with the term *dark matter*.

Explain the need for this term. **2**

Total marks 9

2. The **Hertzsprung-Russell diagram** (**HR diagram**) is one of the most important tools in the study of **stellar evolution**. Amongst other things, it plots **stars' luminosity** against their temperatures.

One can think of the luminosity as being related to the brightness of the star.

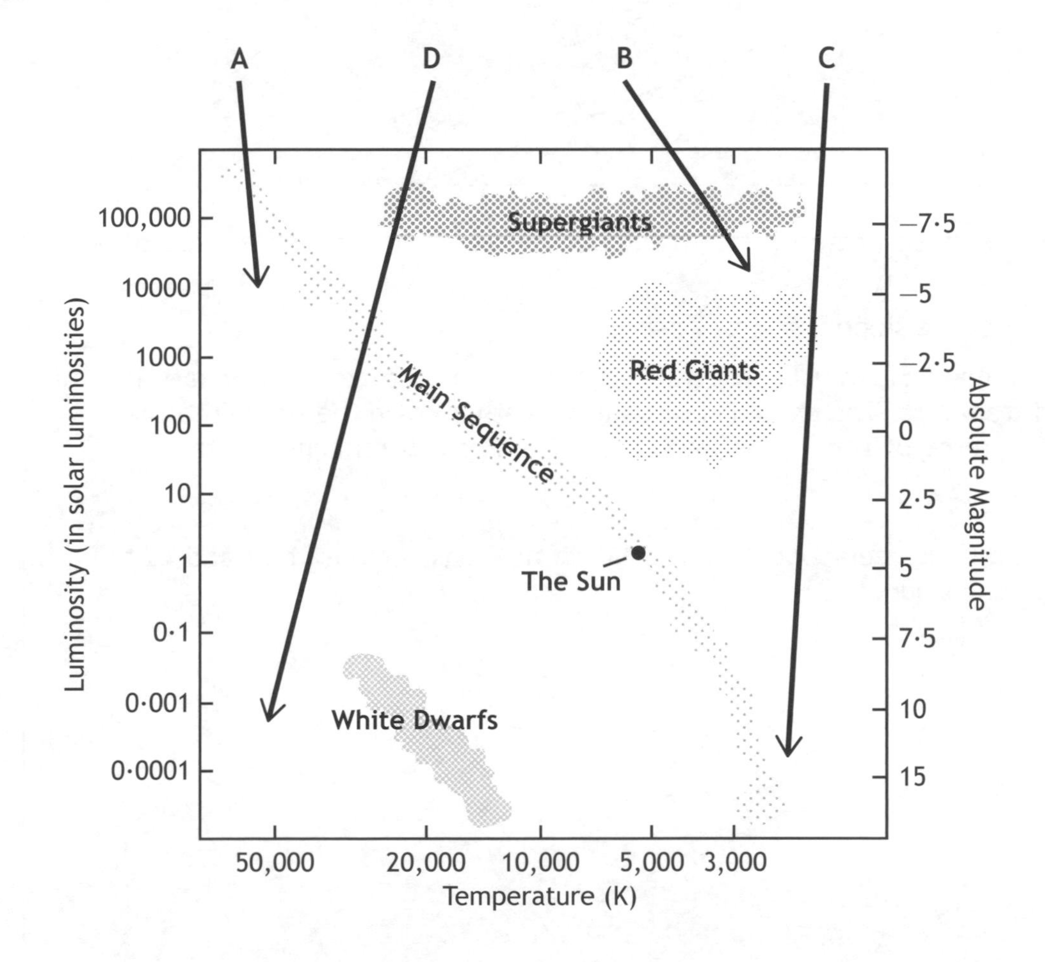

(a) On the diagram, above the letters A, B, C and D, select the words from below that match the area. **2**

Hot, bright Cool, bright Hot, dim Cool, dim

MARKS | DO NOT WRITE IN THIS MARGIN

2. (continued)

(b) Explain how the temperature of each star can be estimated. **2**

Total marks 4

3. A fairground ride consists of rafts which slide down a slope into water.

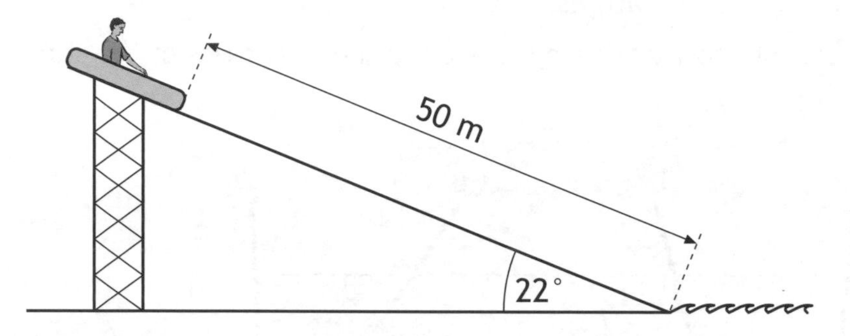

The slope is at an angle of 22° to the horizontal. Each raft has a mass of 8·0kg.

The length of the slope is 50 m.

A child of mass 52 kg sits in a raft at the top of the slope. The raft is released from rest. The child and raft slide together down the slope into the water. The force of friction between the raft and slope remains constant at 180 N.

(a) Calculate the component of weight, in newtons, of the child and raft down the slope. **3**

MARKS | DO NOT WRITE IN THIS MARGIN

3. (continued)

(b) Show by calculation that the acceleration of the child and raft down the slope is 0.67 ms^{-2}. **3**

(c) Calculate the speed of the child and raft at the bottom of the slope. **3**

MARKS | DO NOT WRITE IN THIS MARGIN

3. (continued)

(d) A second child of smaller mass is released from rest in an identical raft at the same starting point. The force of friction is the same as before.

How does the speed of this child and raft at the bottom of the slope compare with the answer to part (c)? **2**

Justify your answer.

Total marks 11

4. The apparatus shown is set up to investigate collisions between two vehicles on a track.

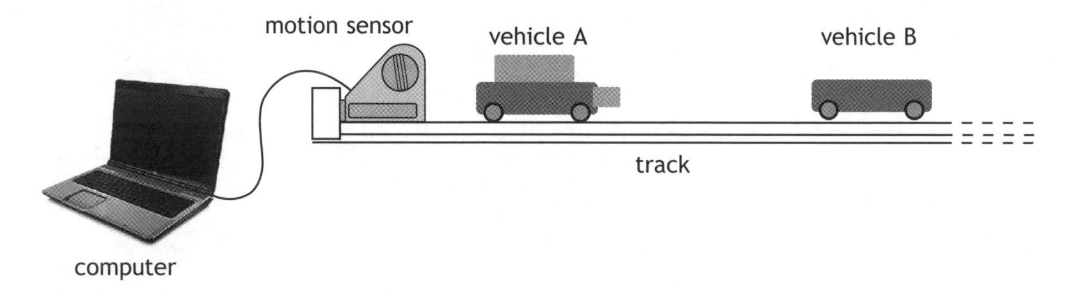

The mass of vehicle A is 0·22 kg and the mass of vehicle B is 0·16 kg.

The effects of friction are negligible.

(a) During one experiment the vehicles collide and stick together. The computer connected to the motion sensor displays the velocity-time graph for vehicle A.

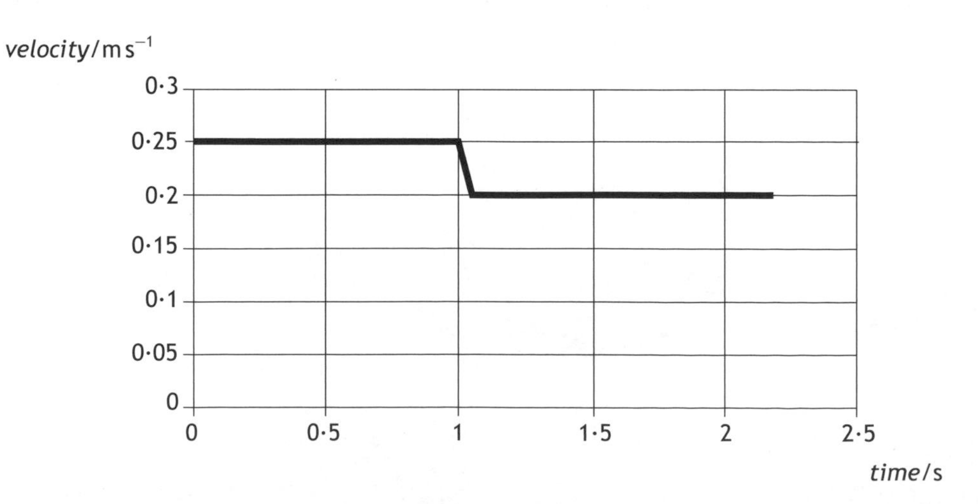

MARKS DO NOT WRITE IN THIS MARGIN

4. (a) (continued)

(i) State the law of conservation of momentum. 1

(ii) Calculate the velocity of vehicle B before the collision. 3

(b) The same apparatus is used to carry out a second experiment.

In this experiment, vehicle B is stationary before the collision.

Vehicle A has the same velocity before the collision as in the first experiment.

After the collision, the two vehicles stick together.

Is their combined velocity less than, equal to, or greater than that in the first collision? 2

Justify your answer.

Total marks 6

MARKS DO NOT WRITE IN THIS MARGIN

5. A student investigates the motion of a ball projected from a launcher.

The launcher is placed on the ground and a ball is fired vertically upwards.

The vertical speed of the ball as it leaves the top of the launcher is $7{\cdot}0\ \text{m s}^{-1}$.

The effects of air resistance can be ignored.

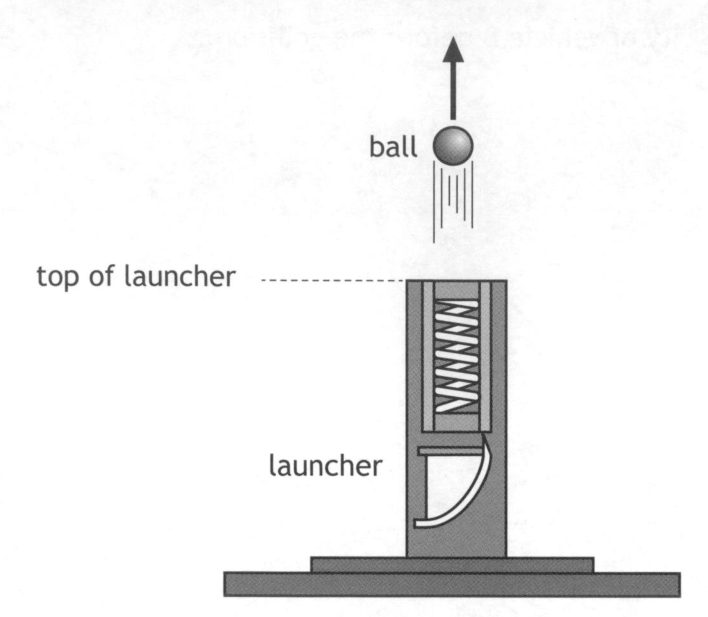

(a) (i) Calculate the maximum height above the top of the launcher reached by the ball. **3**

(ii) Show that the time taken for the ball to reach its maximum height is 0·71 s. **2**

MARKS DO NOT WRITE IN THIS MARGIN

5. (continued)

(b) The student now fixes the launcher to a trolley. The trolley travels horizontally at a constant speed of 1·5 m s^{-1} to the right.

The launcher again fires the ball vertically upwards with a speed of 7·0 m s^{-1}.

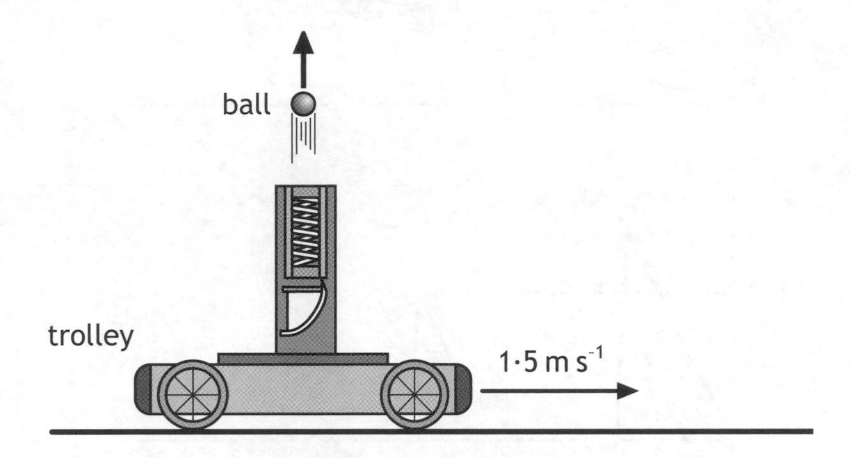

(i) Determine the velocity of the ball after 0·71 s. **1**

(ii) The student asks some friends to predict where the ball will land relative to the moving launcher. They make the following statements.

Statement X: The ball will land behind the launcher.

Statement Y: The ball will land in front of the launcher.

Statement Z: The ball will land on top of the launcher.

Which of the statements is correct? **2**

You must justify your answer.

Total marks 8

MARKS | DO NOT WRITE IN THIS MARGIN

6. A spring loaded trolley is released from the top of a slope. It collides and rebounds at the bottom of the slope.

The velocity – time graph shows the resultant motion of the trolley.

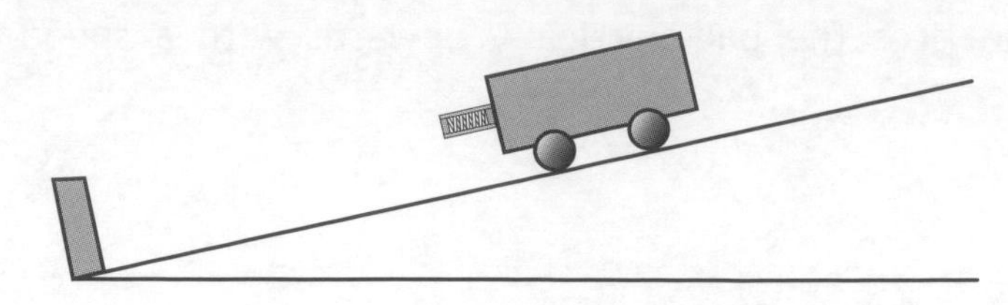

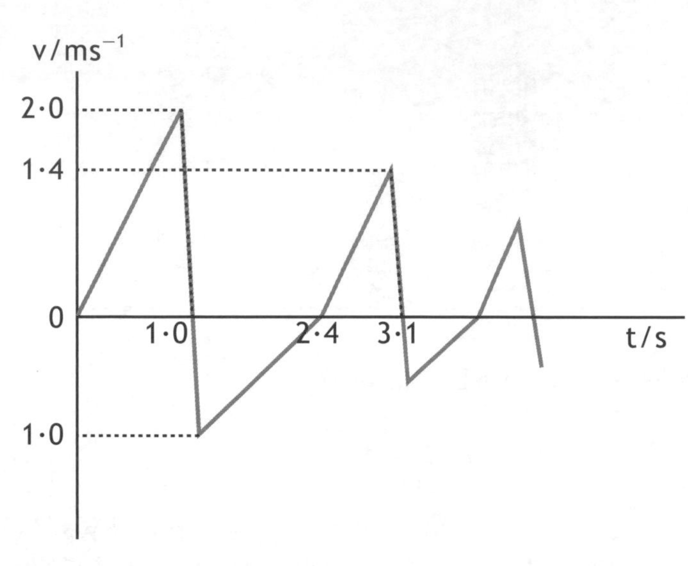

Use your knowledge of Physics to comment on the shape of the graph. **3**

MARKS DO NOT WRITE IN THIS MARGIN

7. (a) An up quark has a charge of $+\frac{2}{3}$e and a down quark a charge of $-\frac{1}{3}$e.

An anti quark has the opposite charge to a quark.

Complete the table for the listed particles. **4**

Particle	Charge	Quark composition
Proton		
Neutron	0	
Pion p^{+}	+e	

(b) State the difference between a composition of a baryon and a meson. **2**

(c) (i) State one difference and one similarity between a particle and its antiparticle. **1**

(ii) Give one example of each. **1**

(iii) What will happen if these particles meet? **1**

Total marks 9

MARKS DO NOT WRITE IN THIS MARGIN

8. The diagram shows a light sensor connected to a voltmeter.

A small lamp is placed in front of the sensor.

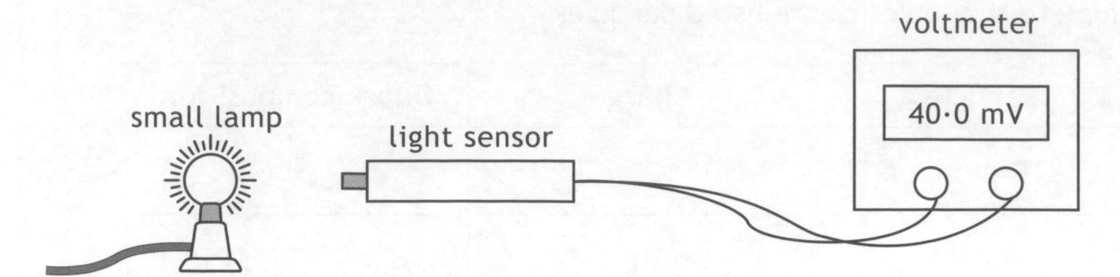

The reading on the voltmeter is 20 mV for each 1·0 mW of power incident on the sensor.

(a) The reading on the voltmeter is 40·0 mV.

The area of the light sensor is $8{\cdot}0 \times 10^{-5}$ m^2.

Calculate the irradiance of light on the sensor. **4**

MARKS | DO NOT WRITE IN THIS MARGIN

8. (continued)

(b) The small lamp is replaced by a different source of light.

Using this new source, a student investigates how irradiance varies with distance.

The results are shown.

Distance/m	0·5	0·7	0·9
Irradiance/W m^{-2}	1·1	0·8	0·6

Can this new source be considered to a point source of light? **2**

Use **all** the data to justify your answer.

Total marks 6

9. You are given the following apparatus.

Ray box

Single slit

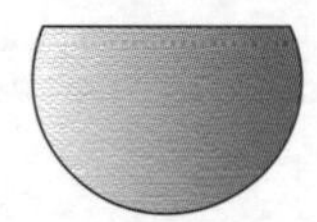

Perspex block

A4 paper, a ruler, a pencil and protractor are also supplied.

Show how you would use this apparatus to find the refractive index of perspex.

You should include:

(a) a clear diagram of the apparatus set up, showing the angles to be measured. **3**

MARKS | DO NOT WRITE IN THIS MARGIN

9. (continued)

(b) a description of the procedure 3

(c) an indication of how to calculate the uncertainties. 2

(d) a graphical method for calculating the value of the refractive index. 3

Total marks 11

MARKS | DO NOT WRITE IN THIS MARGIN

10. (a) Light of frequency $6 \cdot 7 \times 10^{14}$ Hz is produced at the junction of a light emitting diode (LED).

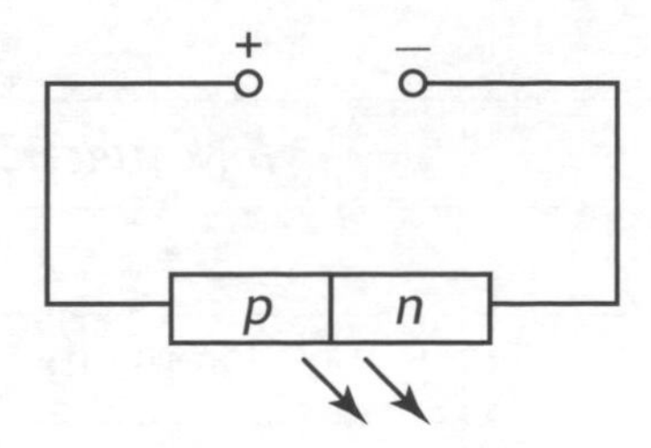

(i) Describe how the movement of charges in a forward biased LED produces light. Your description should include:

electrons, photons, conduction and valence band. **2**

(ii) Calculate the wavelength of the light emitted from the LED. **3**

MARKS DO NOT WRITE IN THIS MARGIN

10. (a) (continued)

(iii) The table below gives the values of the work function for three metals.

Metal	*Work function*/ J
caesium	$3{\cdot}4 \times 10^{-19}$
strontium	$4{\cdot}1 \times 10^{-19}$
magnesium	$5{\cdot}9 \times 10^{-19}$

Light from the LED is now incident on these metals in turn.

Show by calculation which of these metals, if any, release(s) photoelectrons with this light. 4

MARKS | DO NOT WRITE IN THIS MARGIN

10. (continued)

(b) Light from a different LED is passed through a grating as shown below.

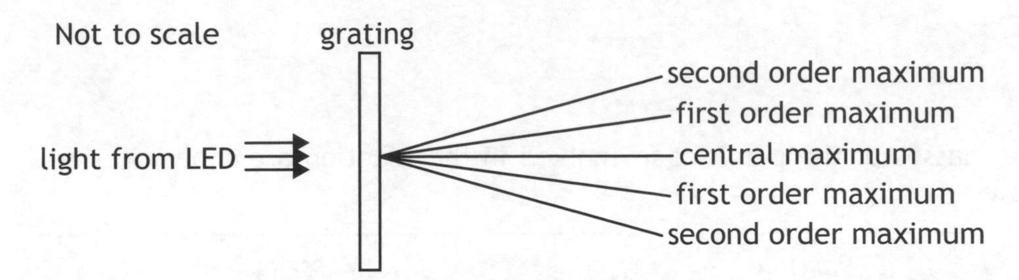

Light from this LED has a wavelength of $6{\cdot}35 \times 10^{-7}$m. The spacing between lines in the grating is $5{\cdot}0 \times 10^{-6}$m.

Calculate the angle between the central maximum and the **second** order maximum. **3**

Total marks 12

11. $^{235}_{92}\text{U} + ^{1}_{0}\text{n} \rightarrow ^{139}_{57}\text{La} + ^{r}_{42}\text{Mo} + 2^{1}_{0}\text{n} + s\,^{0}_{-1}\text{e}$

(a) Determine the numbers represented by the letters *r* and *s* in the above statement. **2**

MARKS DO NOT WRITE IN THIS MARGIN

11. (continued)

(b) Explain why a nuclear fission reaction releases energy. **2**

(c) The masses of the particles invmolved in the reaction are shown in the table.

Particle	*Mass*/kg
${}^{235}_{92}\text{U}$	$390{\cdot}173 \times 10^{-27}$
${}^{139}_{57}\text{La}$	$230{\cdot}584 \times 10^{-27}$
${}^{r}_{42}\text{Mo}$	$157{\cdot}544 \times 10^{-27}$
${}^{1}_{0}\text{n}$	$1{\cdot}675 \times 10^{-27}$
${}^{0}_{-1}\text{e}$	negligible

Calculate the energy released in this reaction. **5**

Total marks 9

MARKS DO NOT WRITE IN THIS MARGIN

12. A student uses a diffraction grating to study red light from a laser and a light emitting diode.

For each, under identical conditions, she directs the light through a diffraction grating onto a screen in a darkened room.

She writes in her report that

Bright and dark areas were produced on the screen for both.

In the case of the laser, the bright areas were clear and well defined.

However for the LED, they appeared broader, slightly further apart and more blurred than expected.

Use your knowledge of Physics to comment on this statement. **3**

13. (a) A supply of e.m.f. 10·0 V and internal resistance r is connected in a circuit as shown in Figure 1.

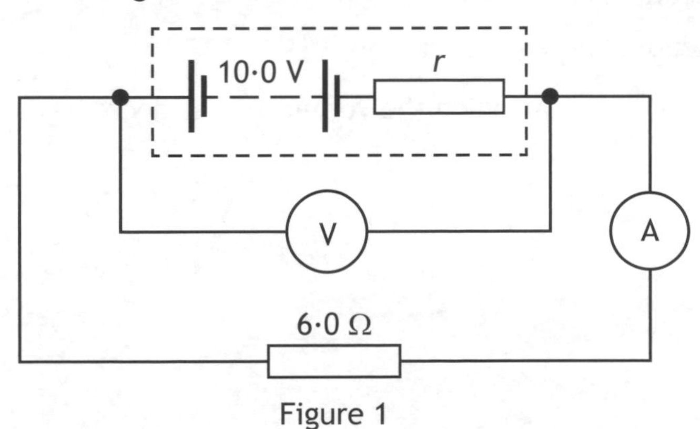

Figure 1

The meters display the following readings.

Reading on ammeter = 1·25 A

Reading on voltmeter = 7·50 V

MARKS | DO NOT WRITE IN THIS MARGIN

13. (a) (continued)

(i) What is meant by an *e.m.f. of 10·0 V*? **1**

(ii) Show that the internal resistance, r, of the supply is 2·0 Ω. **2**

(b) A resistor R is connected to the circuit as shown in Figure 2.

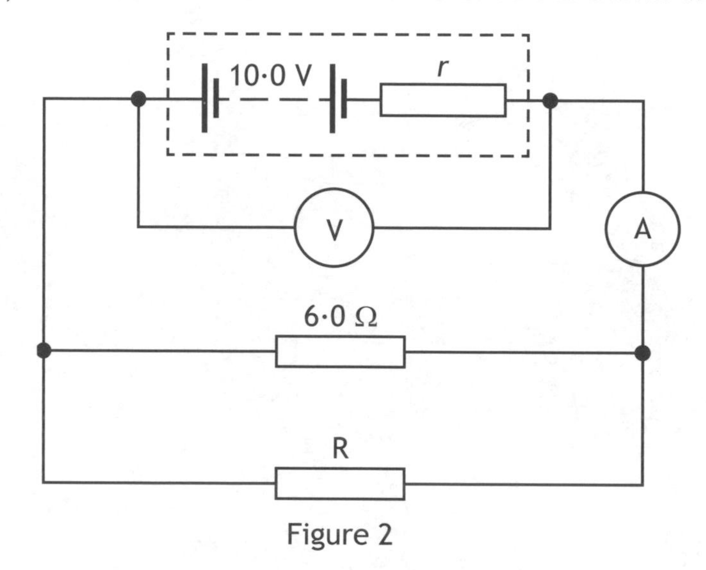

Figure 2

The meters now display the following readings.

Reading on ammeter = 2·0 A

Reading on voltmeter = 6·0 V

(i) Explain why the reading on the voltmeter has decreased. **2**

MARKS DO NOT WRITE IN THIS MARGIN

13. (b) (continued)

(ii) Calculate the resistance of resistor R. 4

Total marks 9

MARKS DO NOT WRITE IN THIS MARGIN

14. A 12 volt battery of negligible internal resistance is connected in a circuit as shown.

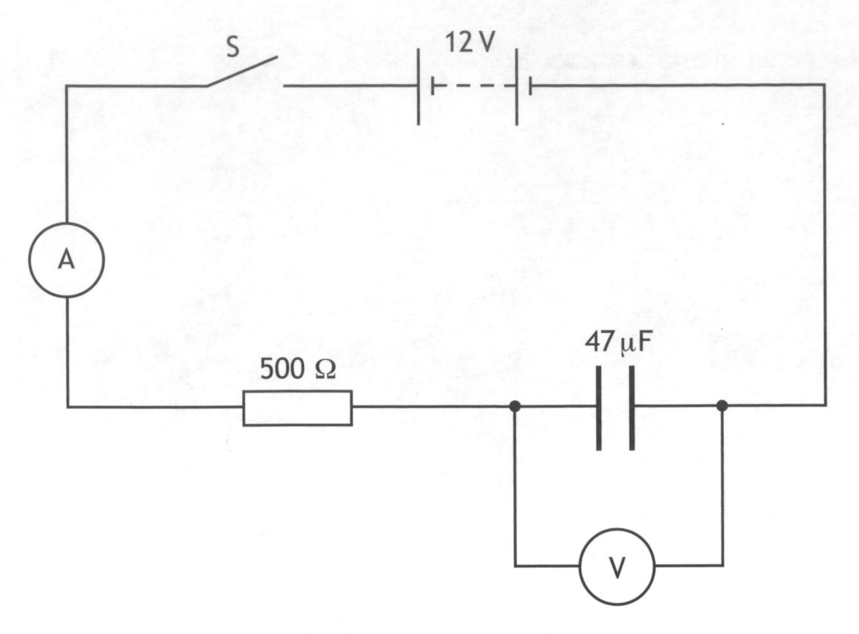

The capacitor is initially uncharged. Switch S is then closed and the capacitor starts to charge.

(a) Sketch a graph of voltage against time from the instant switch S is closed. Numerical values are not required. **1**

MARKS DO NOT WRITE IN THIS MARGIN

14. (continued)

(b) At one instant during the charging of the capacitor the reading on the ammeter is 5·0 mA.

Calculate the reading on the voltmeter at this instant. **4**

(c) Calculate the **maximum** energy stored in the capacitor in this circuit. **3**

MARKS DO NOT WRITE IN THIS MARGIN

14. (continued)

(d) The 500Ω resistor is now replaced with a 2·0 kΩ resistor.

What effect, if any, does this have on the maximum energy stored in the capacitor? **2**

Justify your answer.

Total marks 10

[END OF MODEL PAPER]

ADDITIONAL SPACE FOR ANSWERS AND ROUGH WORK

MARKS | DO NOT WRITE IN THIS MARGIN

ADDITIONAL SPACE FOR ANSWERS AND ROUGH WORK

MARKS | DO NOT WRITE IN THIS MARGIN

HIGHER FOR CfE

Model Paper 3

Whilst this Model Paper has been specially commissioned by Hodder Gibson for use as practice for the Higher (for Curriculum for Excellence) exams, the key reference documents remain the SQA Specimen Paper 2014 and SQA Past Paper 2015.

Physics
Section 1—Questions

Duration — 2 hours and 30 minutes

Instructions for the completion of Section 1 are given on *Page two* of your question and answer booklet.

Record your answers on the answer grid on *Page three* of your question and answer booklet.

Reference may be made to the Data Sheet on *Page two* of this booklet and to the Relationships Sheet.

Before leaving the examination room you must give your question and answer booklet to the Invigilator; if you do not, you may lose all the marks for this paper.

DATA SHEET

COMMON PHYSICAL QUANTITIES

Quantity	*Symbol*	*Value*	*Quantity*	*Symbol*	*Value*
Speed of light in vacuum	c	$3{\cdot}00 \times 10^{8}\,\mathrm{m\,s^{-1}}$	Planck's constant	h	$6{\cdot}63 \times 10^{-34}\,\mathrm{J\,s}$
Magnitude of the charge on an electron	e	$1{\cdot}60 \times 10^{-19}\,\mathrm{C}$	Mass of electron	m_e	$9{\cdot}11 \times 10^{-31}\,\mathrm{kg}$
Universal Constant of Gravitation	G	$6{\cdot}67 \times 10^{-11}\,\mathrm{m^3\,kg^{-1}\,s^{-2}}$	Mass of neutron	m_n	$1{\cdot}675 \times 10^{-27}\,\mathrm{kg}$
Gravitational acceleration on Earth	g	$9{\cdot}8\,\mathrm{m\,s^{-2}}$	Mass of proton	m_p	$1{\cdot}673 \times 10^{-27}\,\mathrm{kg}$
Hubble's constant	H_0	$2{\cdot}3 \times 10^{-18}\,\mathrm{s^{-1}}$			

REFRACTIVE INDICES

The refractive indices refer to sodium light of wavelength 589 nm and to substances at a temperature of 273 K.

Substance	*Refractive index*	*Substance*	*Refractive index*
Diamond	2·42	Water	1·33
Crown glass	1·50	Air	1·00

SPECTRAL LINES

Element	*Wavelength*/nm	*Colour*
Hydrogen	656	Red
	486	Blue-green
	434	Blue-violet
	410	Violet
	397	Ultraviolet
	389	Ultraviolet
Sodium	589	Yellow

Element	*Wavelength*/nm	*Colour*
Cadmium	644	Red
	509	Green
	480	Blue

Lasers

Element	*Wavelength*/nm	*Colour*
Carbon dioxide	9550, 10590	Infrared
Helium-neon	633	Red

PROPERTIES OF SELECTED MATERIALS

Substance	*Density*/$\mathrm{kg\,m^{-3}}$	*Melting Point*/K	*Boiling Point*/K
Aluminium	$2{\cdot}70 \times 10^{3}$	933	2623
Copper	$8{\cdot}96 \times 10^{3}$	1357	2853
Ice	$9{\cdot}20 \times 10^{2}$	273	....
Sea Water	$1{\cdot}02 \times 10^{3}$	264	377
Water	$1{\cdot}00 \times 10^{3}$	273	373
Air	1·29	....	....
Hydrogen	$9{\cdot}0 \times 10^{-2}$	14	20

The gas densities refer to a temperature of 273 K and a pressure of $1{\cdot}01 \times 10^{5}$ Pa.

SECTION 1 — 20 marks

Attempt ALL questions

1. A vehicle is travelling in a straight line. Graphs of velocity and acceleration against time are shown.

   Which pair of graphs could represent the motion of the vehicle?

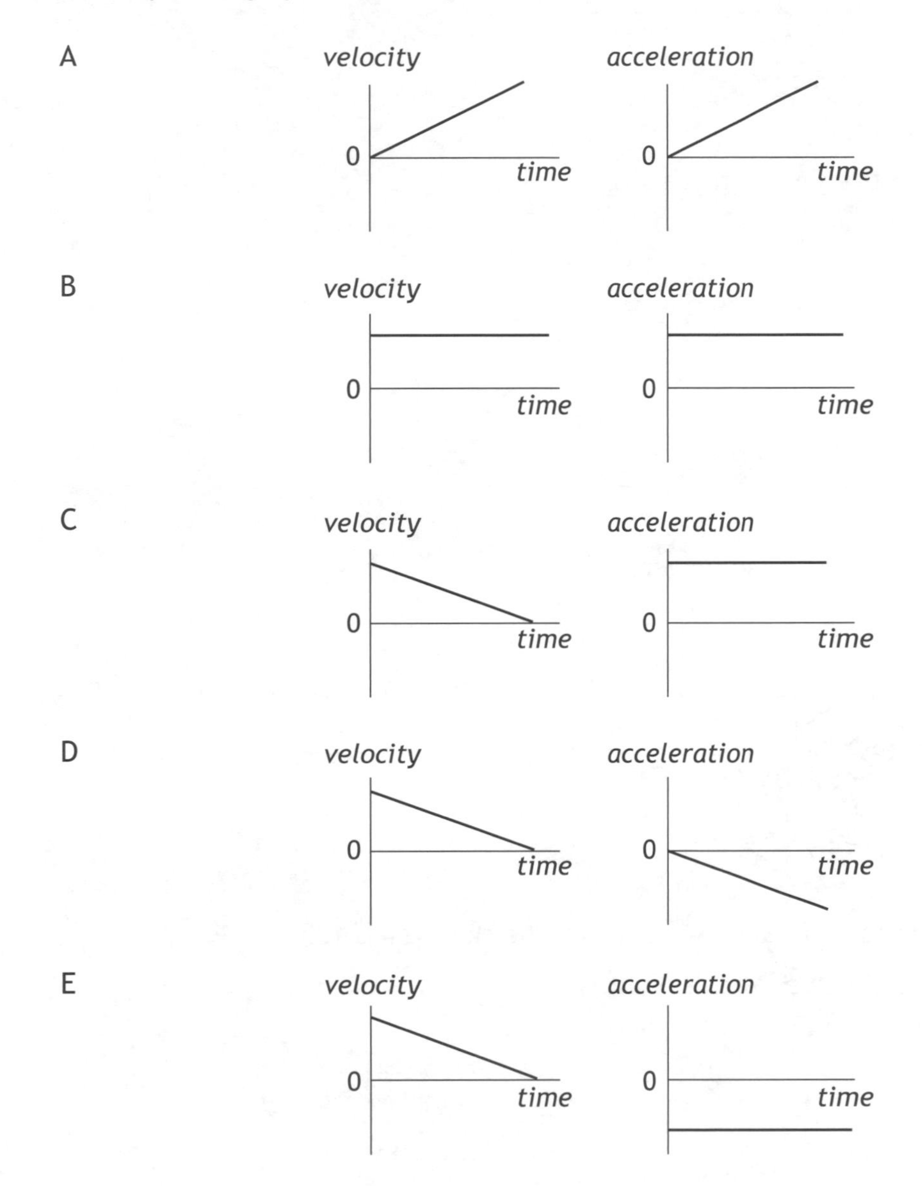

2. A javelin is thrown at 60° to the horizontal with a speed of 20 m s^{-1}.

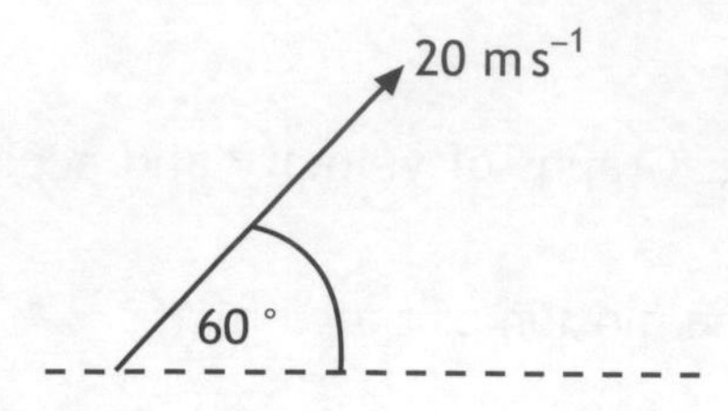

The javelin is in flight for 3.5 s.

Air resistance is negligible.

The horizontal distance the javelin travels is

A 35·0 m

B 60·6 m

C 70·0 m

D 121 m

E 140 m.

3. A skydiver of total mass 85 kg is falling vertically.

At one point during the fall, the air resistance on the skydiver is 135 N.

The acceleration of the skydiver at this point is

A 0·6 m s^{-2}

B 1·6 m s^{-2}

C 6·2 m s^{-2}

D 8·2 m s^{-2}

E 13·8 m s^{-2}.

4. The graph shows the force which acts on an object over a time interval of 8 seconds.

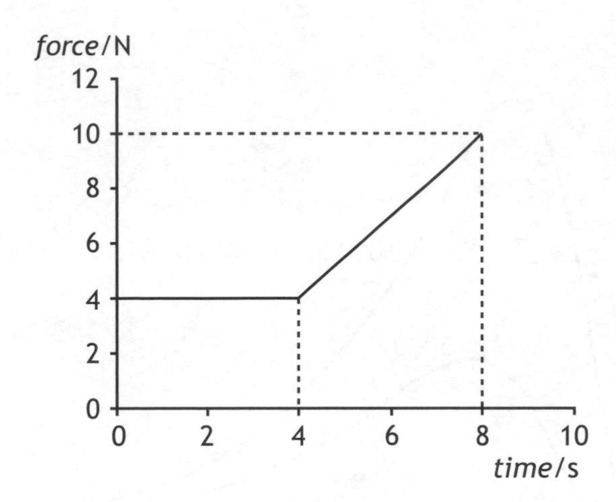

The momentum gained by the object during this 8 seconds is

A 12 kg m s^{-1}

B 32 kg m s^{-1}

C 44 kg m s^{-1}

D 52 kg m s^{-1}

E 72 kg m s^{-1}.

5.

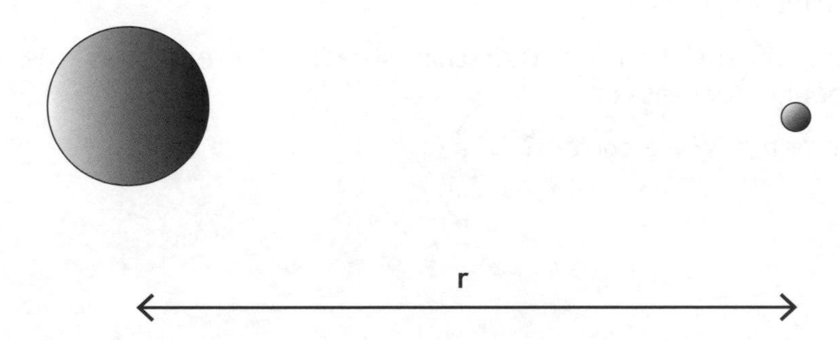

The force acting on a mass a distance r from the earth's centre is 500 N.

The force acting on the same mass at a distance of 4r from the earth's centre will be:

A 125 N

B 0

C 100 N

D 31·3 N

E 20 N

6. The graph shows how the energy emitted per second from the surface of a hot object varies with the wavelength, λ, of the emitted radiation at different temperatures.

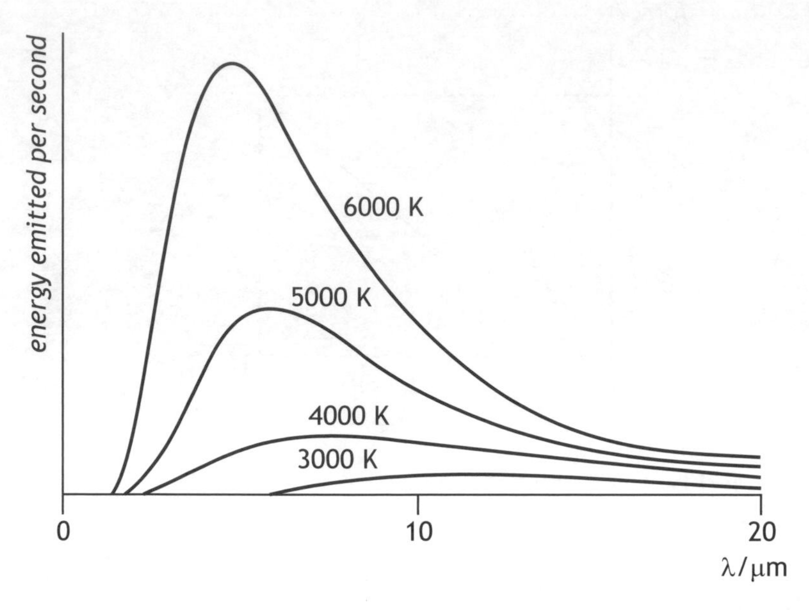

A student makes the following statements based on the information shown in the graph.

I As the temperature of the object increases, the total energy emitted per second decreases.

II As the temperature of the object increases, the peak wavelength of the emitted radiation decreases.

III The frequency of the emitted radiation steadily increases as the emitted energy per second decreases.

Which of the statements is/are correct?

A I only

B II only

C III only

D I and II only

E II and III only

7. A physicist, on earth, observes a spacecraft travelling at 0·8c.

Which of the following will describe the:

(a) observed time interval on the spacecraft compared to that on earth

(b) observed length of spacecraft compared to it being stationary on earth.

	Time Interval	*Length*
A	Slower	Shorter
B	Faster	Shorter
C	No difference	No difference
D	Slower	Longer
E	Faster	Longer

8. Which of the following statements are correct:

I A baryon consists of 3 quarks

II Leptons consist of 3 quarks

III Mesons consist of 2 quarks

A I only

B II only

C I and II

D I and III

E I, II and III.

9. Light travels from air into glass.

Which row in the table describes what happens to the speed, frequency and wavelength of the light?

	Speed	*Frequency*	*Wavelength*
A	increases	stays constant	increases
B	increases	decreases	stays constant
C	stays constant	decreases	decreases
D	decreases	decreases	stays constant
E	decreases	stays constant	decreases

10. Two identical loudspeakers, L1 and L2, are connected to a signal generator as shown.

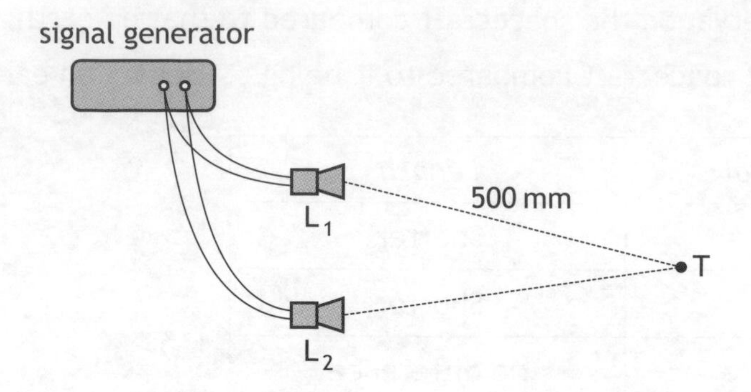

An interference pattern is produced.

A minimum is detected at point T.

The wavelength of the sound is 40 mm.

The distance from L_1 to T is 500 mm.

The distance from L_2 to T is

A 450 mm

B 460 mm

C 470 mm

D 480 mm

E 490 mm.

11. A proton is fired though a magnetic field as shown.

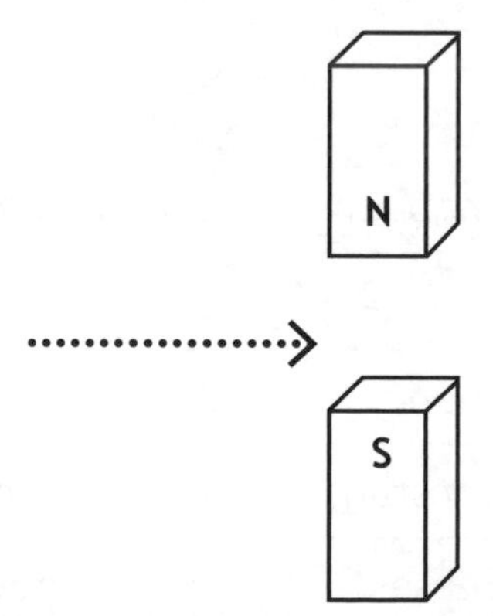

On entering the field, the direction of the force acting on the proton will be:

A Upwards

B Downwards

C To the left.

D Into the page

E Out of the page

12. For the nuclear decay shown, which row of the table gives the current values of x, y and z?

$$^{214}_{x}\text{Pb} \longrightarrow {}^{y}_{83}\text{Bi} + {}^{0}_{z}\text{e}$$

	x	y	z
A	85	214	2
B	84	214	1
C	83	210	4
D	82	214	−1
E	82	210	−1

13. Electromagnetic radiation of frequency $9{\cdot}0 \times 10^{14}$ Hz is incident on a clean metal surface.

The work function of the metal is $5{\cdot}0 \times 10^{-19}$ J.

The maximum kinetic energy of a photoelectron released from the metal surface is

A $1{\cdot}0 \times 10^{-19}$ J

B $4{\cdot}0 \times 10^{-19}$ J

C $5{\cdot}0 \times 10^{-19}$ J

D $6{\cdot}0 \times 10^{-19}$ J

E $9{\cdot}0 \times 10^{-19}$ J.

14. Part of the energy level diagram for an atom is shown.

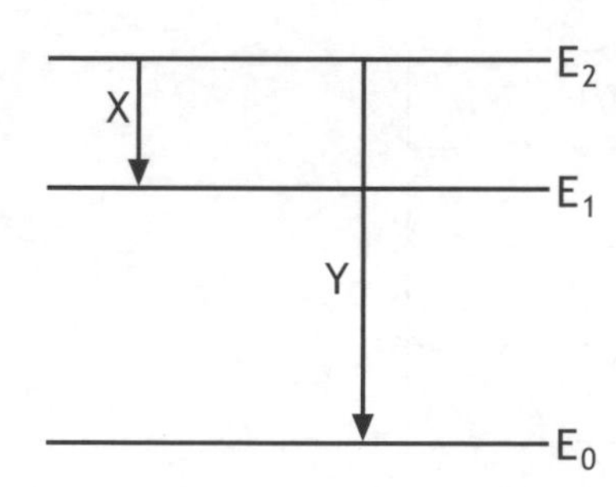

X and Y represent two possible electron transitions.

14. (continued)

Which of the following statements is/are correct?

I Transition Y produces photons of higher frequency than transition X.

II Transition X produces photons of longer wavelength than transition Y.

III When an electron is in the energy level E_0, the atom is ionised.

A I only

B I and II only

C I and III only

D II and III only

E I, II and III

15. The diagram shows the trace on an oscilloscope when an alternating voltage is applied to its input.

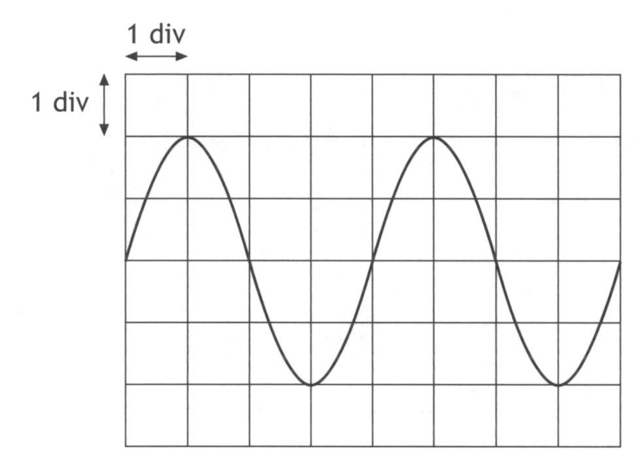

The timebase is set at 5 ms/div and the Y-gain is set at 10 V/div.

Which row in the table gives the peak voltage and the frequency of the signal?

	Peak voltage/V	*Frequency*/Hz
A	7·1	20
B	14	50
C	20	20
D	20	50
E	40	50

16. The element of an electric kettle has a resistance of 30 Ω. The kettle is connected to a mains supply. The r.m.s. voltage of this supply is 230V. The peak value of the current in the kettle is

A 0·13 A

B 0·18 A

C 5·4 A

D 7·7 A

E 10·8 A.

17. In the following circuit, the battery has an e.m.f. of 8·0 V and an internal resistance of 0·20 Ω.

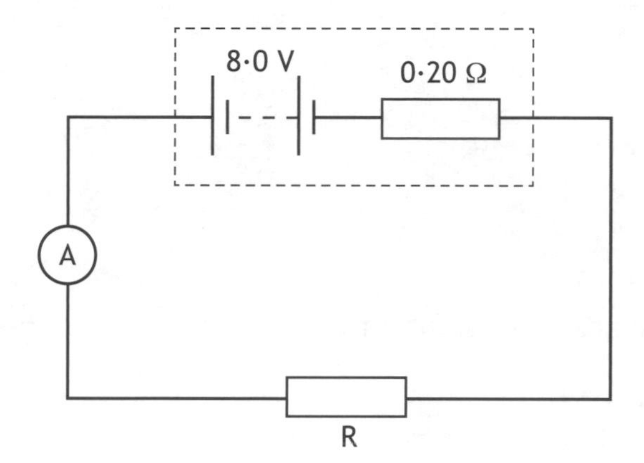

The reading on the ammeter is 4·0 A.
The resistance of R is

A 0·5 Ω

B 1·8 Ω

C 2·0 Ω

D 2·2 Ω

E 6·4 Ω

18. Four resistors, each of resistance 20 Ω, are connected to a 60 V supply of negligible internal resistance, as shown.

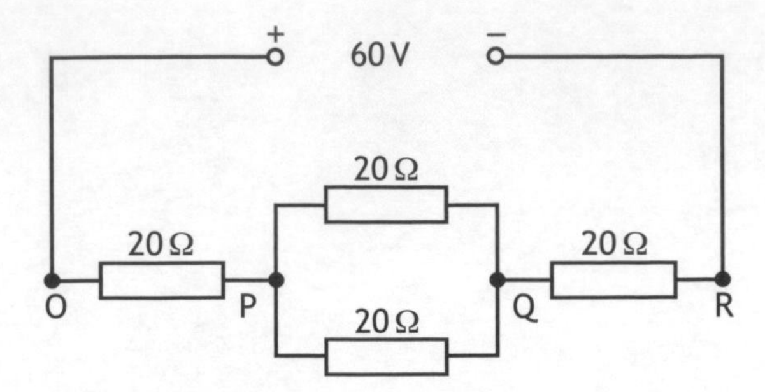

The potential difference across QR is

A 12 V

B 15 V

C 20 V

D 24 V

E 30 V.

19. A supply with a sinusoidally alternating output of 6·0 V r.m.s. is connected to a 3·0 Ω resistor.

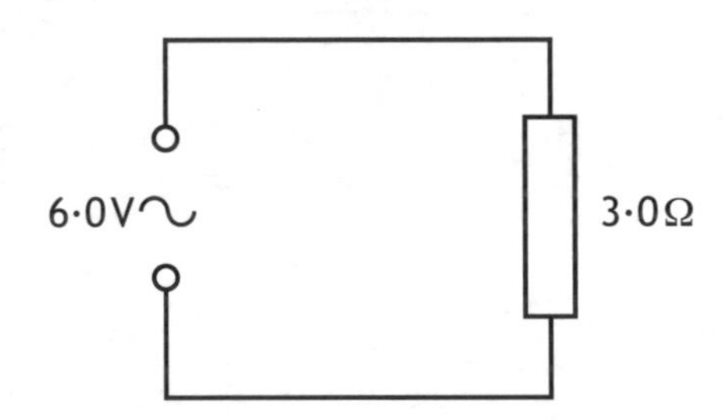

Which row in the following table shows the peak voltage across the resistor and the peak current in the circuit?

	Peak voltage/V	*Peak current*/A
A	$6\sqrt{2}$	$2\sqrt{2}$
B	$6\sqrt{2}$	2
C	6	2
D	$6\sqrt{2}$	$\frac{1}{2\sqrt{2}}$
E	6	$2\sqrt{2}$

20. The letters **X**, **Y** and **Z** represent three missing words from the following passage.

Materials can be divided into three broad categories according to their electrical resistance.

......X...... *have a very high resistance.*

......Y...... *have a high resistance in their pure form but when small amounts of certain impurities are added, the resistance decreases.*

......Z...... *have a very low resistance.*

Which row in the table shows the missing words?

	X	Y	Z
A	conductors	insulators	semi-conductors
B	semi-conductors	insulators	conductors
C	insulators	semi-conductors	conductors
D	conductors	semi-conductors	insulators
E	insulators	conductors	semi-conductors

[END OF SECTION 1. NOW ATTEMPT THE QUESTIONS IN SECTION 2 OF YOUR QUESTION AND ANSWER BOOKLET]

National
Qualifications
MODEL PAPER 3

Physics
Relationships Sheet

Date — Not applicable

Relationships required for Physics Higher

$d = \bar{v}t$

$s = \bar{v}t$

$v = u + at$

$s = ut + \frac{1}{2}at^2$

$v^2 = u^2 + 2as$

$s = \frac{1}{2}(u + v)t$

$W = mg$

$F = ma$

$E_W = Fd$

$E_p = mgh$

$E_k = \frac{1}{2}mv^2$

$P = \frac{E}{t}$

$p = mv$

$Ft = mv - mu$

$F = G\frac{m_1 m_2}{r^2}$

$t' = \frac{t}{\sqrt{1 - \left(\frac{v}{c}\right)^2}}$

$l' = l\sqrt{1 - \left(\frac{v}{c}\right)^2}$

$f_o = f_s\left(\frac{v}{v \pm v_s}\right)$

$z = \frac{\lambda_{observed} - \lambda_{rest}}{\lambda_{rest}}$

$z = \frac{v}{c}$

$v = H_0 d$

$W = QV$

$E = mc^2$

$E = hf$

$E_k = hf - hf_0$

$E_2 - E_1 = hf$

$T = \frac{1}{f}$

$v = f\lambda$

$d\sin\theta = m\lambda$

$n = \frac{\sin\theta_1}{\sin\theta_2}$

$\frac{\sin\theta_1}{\sin\theta_2} = \frac{\lambda_1}{\lambda_2} = \frac{v_1}{v_2}$

$\sin\theta_c = \frac{1}{n}$

$I = \frac{k}{d^2}$

$I = \frac{P}{A}$

path difference = $m\lambda$ or $\left(m + \frac{1}{2}\right)\lambda$ where $m = 0, 1, 2 \ldots$

random uncertainty = $\frac{\text{max. value} - \text{min. value}}{\text{number of values}}$

$V_{peak} = \sqrt{2}V_{rms}$

$I_{peak} = \sqrt{2}I_{rms}$

$Q = It$

$V = IR$

$P = IV = I^2R = \frac{V^2}{R}$

$R_T = R_1 + R_2 + \ldots$

$\frac{1}{R_T} = \frac{1}{R_1} + \frac{1}{R_2} + \ldots$

$E = V + Ir$

$V_1 = \left(\frac{R_1}{R_1 + R_2}\right)V_s$

$\frac{V_1}{V_2} = \frac{R_1}{R_2}$

$C = \frac{Q}{V}$

$E = \frac{1}{2}QV = \frac{1}{2}CV^2 = \frac{1}{2}\frac{Q^2}{C}$

Additional Relationships

Circle

circumference = $2\pi r$

area = πr^2

Sphere

area = $4\pi r^2$

volume = $\frac{4}{3}\pi r^3$

Trigonometry

$$\sin\Theta = \frac{\text{opposite}}{\text{hypotenuse}}$$

$$\cos\Theta = \frac{\text{adjacent}}{\text{hypotenuse}}$$

$$\tan\Theta = \frac{\text{opposite}}{\text{adjacent}}$$

$$\sin^2\Theta + \cos^2\Theta = 1$$

Electron Arrangements of Elements

Key

Atomic number Symbol Electron arrangement Name

Transition Elements

Group 1 (1)	Group 2 (2)	(3)	(4)	(5)	(6)	(7)	(8)	(9)	(10)	(11)	(12)	Group 3 (13)	Group 4 (14)	Group 5 (15)	Group 6 (16)	Group 7 (17)	Group 0 (18)
1 **H** 1 Hydrogen																	2 **He** 2 Helium
3 **Li** 2,1 Lithium	4 **Be** 2,2 Beryllium											5 **B** 2,3 Boron	6 **C** 2,4 Carbon	7 **N** 2,5 Nitrogen	8 **O** 2,6 Oxygen	9 **F** 2,7 Fluorine	10 **Ne** 2,8 Neon
11 **Na** 2,8,1 Sodium	12 **Mg** 2,8,2 Magnesium											13 **Al** 2,8,3 Aluminium	14 **Si** 2,8,4 Silicon	15 **P** 2,8,5 Phosphorus	16 **S** 2,8,6 Sulfur	17 **Cl** 2,8,7 Chlorine	18 **Ar** 2,8,8 Argon
19 **K** 2,8,8,1 Potassium	20 **Ca** 2,8,8,2 Calcium	21 **Sc** 2,8,9,2 Scandium	22 **Ti** 2,8,10,2 Titanium	23 **V** 2,8,11,2 Vanadium	24 **Cr** 2,8,13,1 Chromium	25 **Mn** 2,8,13,2 Manganese	26 **Fe** 2,8,14,2 Iron	27 **Co** 2,8,15,2 Cobalt	28 **Ni** 2,8,16,2 Nickel	29 **Cu** 2,8,18,1 Copper	30 **Zn** 2,8,18,2 Zinc	31 **Ga** 2,8,18,3 Gallium	32 **Ge** 2,8,18,4 Germanium	33 **As** 2,8,18,5 Arsenic	34 **Se** 2,8,18,6 Selenium	35 **Br** 2,8,18,7 Bromine	36 **Kr** 2,8,18,8 Krypton
37 **Rb** 2,8,18,8,1 Rubidium	38 **Sr** 2,8,18,8,2 Strontium	39 **Y** 2,8,18,9,2 Yttrium	40 **Zr** 2,8,18, 10,2 Zirconium	41 **Nb** 2,8,18, 12,1 Niobium	42 **Mo** 2,8,18,13, 1 Molybdenum	43 **Tc** 2,8,18,13, 2 Technetium	44 **Ru** 2,8,18,15, 1 Ruthenium	45 **Rh** 2,8,18,16, 1 Rhodium	46 **Pd** 2,8,18, 18,0 Palladium	47 **Ag** 2,8,18, 18,1 Silver	48 **Cd** 2,8,18, 18,2 Cadmium	49 **In** 2,8,18, 18,3 Indium	50 **Sn** 2,8,18, 18,4 Tin	51 **Sb** 2,8,18, 18,5 Antimony	52 **Te** 2,8,18, 18,6 Tellurium	53 **I** 2,8,18, 18,7 Iodine	54 **Xe** 2,8,18, 18,8 Xenon
55 **Cs** 2,8,18,18, 8,1 Caesium	56 **Ba** 2,8,18,18, 8,2 Barium	57 **La** 2,8,18,18, 9,2 Lanthanum	72 **Hf** 2,8,18,32, 10,2 Hafnium	73 **Ta** 2,8,18, 32,11,2 Tantalum	74 **W** 2,8,18,32, 12,2 Tungsten	75 **Re** 2,8,18,32, 13,2 Rhenium	76 **Os** 2,8,18,32, 14,2 Osmium	77 **Ir** 2,8,18,32, 15,2 Iridium	78 **Pt** 2,8,18,32, 17,1 Platinum	79 **Au** 2,8,18, 32,18,1 Gold	80 **Hg** 2,8,18, 32,18,2 Mercury	81 **Tl** 2,8,18, 32,18,3 Thallium	82 **Pb** 2,8,18, 32,18,4 Lead	83 **Bi** 2,8,18, 32,18,5 Bismuth	84 **Po** 2,8,18, 32,18,6 Polonium	85 **At** 2,8,18, 32,18,7 Astatine	86 **Rn** 2,8,18, 32,18,8 Radon
87 **Fr** 2,8,18,32, 18,8,1 Francium	88 **Ra** 2,8,18,32, 18,8,2 Radium	89 **Ac** 2,8,18,32, 18,9,2 Actinium	104 **Rf** 2,8,18,32, 32,10,2 Rutherfordium	105 **Db** 2,8,18,32, 32,11,2 Dubnium	106 **Sg** 2,8,18,32, 32,12,2 Seaborgium	107 **Bh** 2,8,18,32, 32,13,2 Bohrium	108 **Hs** 2,8,18,32, 32,14,2 Hassium	109 **Mt** 2,8,18,32, 32,15,2 Meitnerium	110 **Ds** 2,8,18,32, 32,17,1 Darmstadtium	111 **Rg** 2,8,18,32, 32,18,1 Roentgenium	112 **Cn** 2,8,18,32, 32,18,2 Copernicium						

Lanthanides	57 **La** 2,8,18, 18,9,2 Lanthanum	58 **Ce** 2,8,18, 20,8,2 Cerium	59 **Pr** 2,8,18,21, 8,2 Praseodymium	60 **Nd** 2,8,18,22, 8,2 Neodymium	61 **Pm** 2,8,18,23, 8,2 Promethium	62 **Sm** 2,8,18,24, 8,2 Samarium	63 **Eu** 2,8,18,25, 8,2 Europium	64 **Gd** 2,8,18,25, 9,2 Gadolinium	65 **Tb** 2,8,18,27, 8,2 Terbium	66 **Dy** 2,8,18,28, 8,2 Dysprosium	67 **Ho** 2,8,18,29, 8,2 Holmium	68 **Er** 2,8,18,30, 8,2 Erbium	69 **Tm** 2,8,18,31, 8,2 Thulium	70 **Yb** 2,8,18,32, 8,2 Ytterbium	71 **Lu** 2,8,18,32, 9,2 Lutetium
Actinides	89 **Ac** 2,8,18,32, 18,9,2 Actinium	90 **Th** 2,8,18,32, 18,10,2 Thorium	91 **Pa** 2,8,18,32, 20,9,2 Protactinium	92 **U** 2,8,18,32, 21,9,2 Uranium	93 **Np** 2,8,18,32, 22,9,2 Neptunium	94 **Pu** 2,8,18,32, 24,8,2 Plutonium	95 **Am** 2,8,18,32, 25,8,2 Americium	96 **Cm** 2,8,18,32, 25,9,2 Curium	97 **Bk** 2,8,18,32, 27,8,2 Berkelium	98 **Cf** 2,8,18,32, 28,8,2 Californium	99 **Es** 2,8,18,32, 29,8,2 Einsteinium	100 **Fm** 2,8,18,32, 30,8,2 Fermium	101 **Md** 2,8,18,32, 31,8,2 Mendelevium	102 **No** 2,8,18,32, 32,8,2 Nobelium	103 **Lr** 2,8,18,32, 32,9,2 Lawrencium

Physics
Section 1 — Answer Grid and Section 2

Duration — 2 hours and 30 minutes

Fill in these boxes and read what is printed below.

Full name of centre | Town

Forename(s) | Surname | Number of seat

Date of birth
Day Month Year
D D M M Y Y

Scottish candidate number

Total marks — 130

SECTION 1 — 20 marks

Attempt ALL questions.

Instructions for the completion of Section 1 are given on *Page two*.

SECTION 2 — 110 marks

Attempt ALL questions.

Reference may be made to the Data Sheet on *Page two* of the question paper and to the Relationship Sheet.

Write your answers clearly in the spaces provided in this booklet. Additional space for answers and rough work is provided at the end of this booklet. If you use this space you must clearly identify the question number you are attempting. Any rough work must be written in this booklet. You should score through your rough work when you have written your final copy.

Use **blue** or **black** ink.

Care should be taken to give an appropriate number of significant figures in the final answers to calculations.

Before leaving the examination room you must give this booklet to the Invigilator; if you do not you may lose all the marks for this paper.

HODDER GIBSON
LEARN MORE

SECTION 1 — 20 marks

The questions for Section 1 are contained in the booklet Physics Section 1 – Questions.
Read these and record your answers on the answer grid on *Page three* opposite.
Do **NOT** use gel pens.

1. The answer to each question is **either** A, B, C, D or E. Decide what your answer is, then fill in the appropriate bubble (see sample question below).
2. There is **only one correct** answer to each question.
3. Any rough working should be done on the additional space for answers and rough work at the end of this booklet.

Sample Question

The energy unit measured by the electricity meter in your home is the:

A ampere
B kilowatt-hour
C watt
D coulomb
E volt.

The correct answer is **B**—kilowatt-hour. The answer **B** bubble has been clearly filled in (see below).

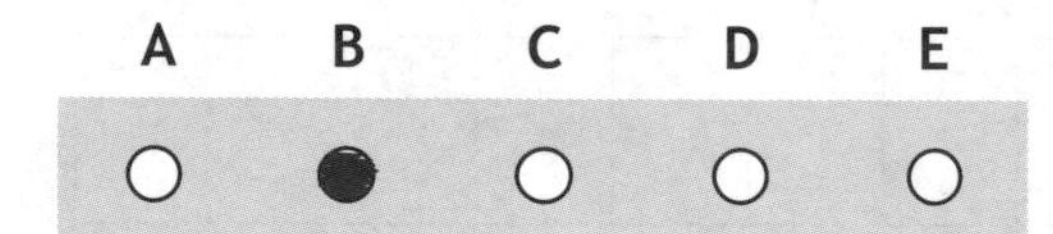

Changing an answer

If you decide to change your answer, cancel your first answer by putting a cross through it (see below) and fill in the answer you want. The answer below has been changed to **D**.

A B C D E

If you then decide to change back to an answer you have already scored out, put a tick (✓) to the **right** of the answer you want, as shown below:

A B C D E

or

A B C D E

SECTION 1 — Answer Grid

	A	B	C	D	E
1	○	○	○	○	○
2	○	○	○	○	○
3	○	○	○	○	○
4	○	○	○	○	○
5	○	○	○	○	○
6	○	○	○	○	○
7	○	○	○	○	○
8	○	○	○	○	○
9	○	○	○	○	○
10	○	○	○	○	○
11	○	○	○	○	○
12	○	○	○	○	○
13	○	○	○	○	○
14	○	○	○	○	○
15	○	○	○	○	○
16	○	○	○	○	○
17	○	○	○	○	○
18	○	○	○	○	○
19	○	○	○	○	○
20	○	○	○	○	○

MARKS | DO NOT WRITE IN THIS MARGIN

SECTION 2 — 110 marks

Attempt ALL questions

1. (a) Light of wavelength 586nm from a distant galaxy appears to be red shifted by 12 nm.

   Calculate the recessional velocity of the galaxy. **3**

   (b) The table below gives the distance to some of our closest galaxies and their recessional speed.

Galaxy	Distance for Milky Way ($\times 10^6$ light years)	Recessional Speed (kms^{-1})
Milky Way	0	0
Virgo	80	1200
Perseus	350	5400
Hercules	650	10000
Ursa Major	1000	15000

   (i) Use a non-graphical method to test whether or not this data agrees with Hubble's Law. **3**

MARKS | DO NOT WRITE IN THIS MARGIN

1. **(b)** (continued)

(ii) Give one reason why the reliability of the data in part (b)(i) is in doubt. **1**

(c) Estimate the age, in years, of the universe using the value of the Hubble constant H_0 given in the data sheet. **3**

Total marks 10

2. A van of mass 2600 kg moves down a slope which is inclined at 12° to the horizontal as shown.

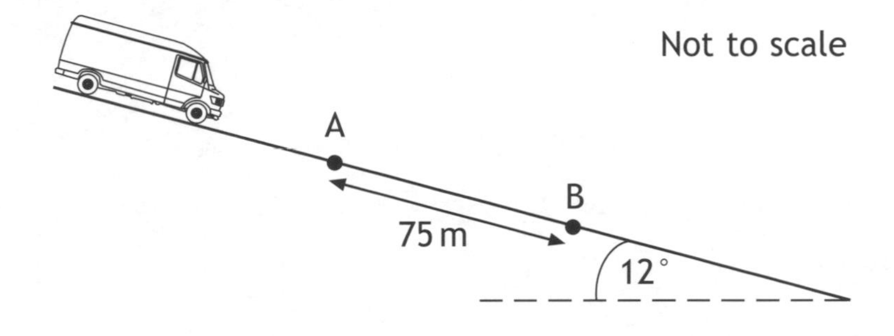

(a) Calculate the component of the van's weight parallel to the slope. **3**

MARKS | DO NOT WRITE IN THIS MARGIN

2. (continued)

(b) A constant frictional force of 1400 N acts on the van as it moves down the slope.

Calculate the acceleration of the van. **3**

(c) The speed of the van as it passes point **A** is $5{\cdot}0\,\text{ms}^{-1}$.

Point **B** is 75 m further down the slope.

Calculate the kinetic energy of the van at **B**. **5**

Total marks 11

MARKS DO NOT WRITE IN THIS MARGIN

3. Golf clubs are tested to ensure they meet certain standards.

(a) In one test, a securely held clubhead is hit by a small steel pendulum. The time of contact between the clubhead and the pendulum is recorded.

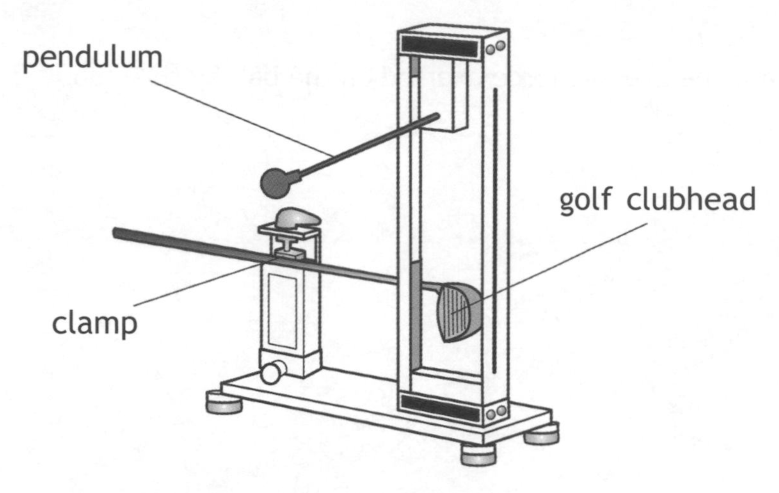

The experiment is repeated several times.

The results are shown.

248 μs   259 μs   251 μs   263 μs   254 μs

(i) Calculate:

(A) the mean contact time between the clubhead and the pendulum; **1**

(B) the approximate absolute random uncertainty in this value. **2**

(ii) In this test, the standard required is that the maximum value of the mean contact time must not be greater than 257 μs.

Does the club meet this standard? **2**

You must justify your answer.

MARKS | DO NOT WRITE IN THIS MARGIN

3. (continued)

(b) In another test, a machine uses a club to hit a stationary golf ball.

The mass of the ball is 4.5×10^{-2} kg. The ball leaves the club with a speed of 50.0 ms^{-1}. The time of contact between the club and ball is 450 μs.

(i) Calculate the average force exerted on the ball by the club. **3**

(ii) The test is repeated using a different club and an identical ball. The machine applies the same average force on the ball but with a longer contact time.

What effect, if any, does this have on the speed of the ball as it leaves the club?

Justify your answer. **2**

Total marks 10

MARKS | DO NOT WRITE IN THIS MARGIN

4. To test the braking system of cars, a test track is set up as shown.

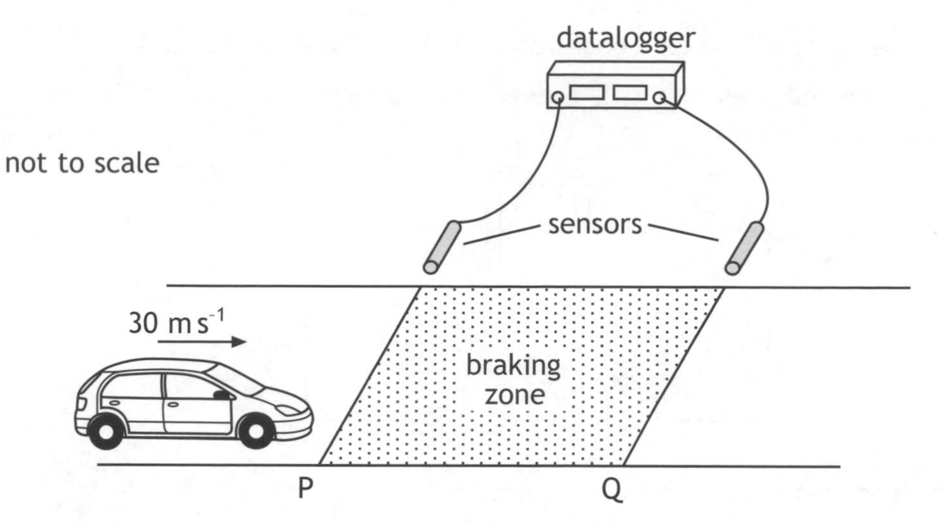

The sensors are connected to a datalogger which records the speed of a car at both P and Q.

A car is driven at a constant speed of 30 m s^{-1} until it reaches the start of the braking zone at P. The brakes are then applied.

(a) In one test, the datalogger records the speed at P as 30 m s^{-1} and the speed at Q as 12 m s^{-1}. The car slows down at a constant rate of 9.0 m s^{-2} between P and Q.

Calculate the length of the braking zone. **3**

(b) The test is repeated. The same car is used but now with passengers in the car. The speed at P is again recorded as 30 m s^{-1}.

The same braking force is applied to the car as in part (a).

How does the speed of the car at Q compare with its speed at Q in part (a)? **2**

Justify your answer.

MARKS DO NOT WRITE IN THIS MARGIN

4. (continued)

(c) The brake lights of the car consist of a number of very bright LEDs.

An LED from the brake lights is forward biased by connecting it to a 12V car battery as shown.

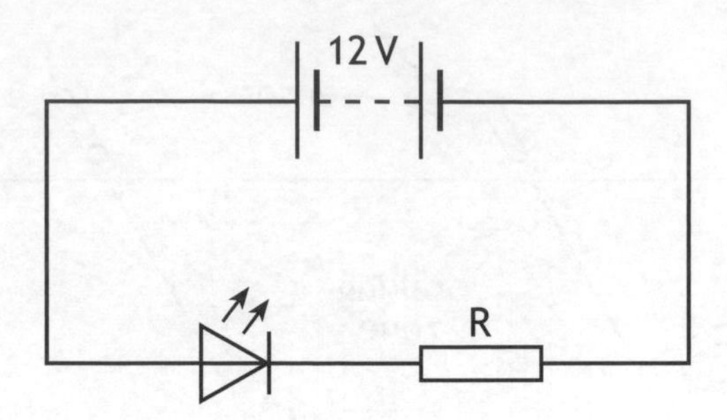

The battery has negligible internal resistance.

(i) Explain, in terms of valence and conduction bands, how the LED can emit light of different colours. **3**

(ii) The LED is operating at its rated values of 5·0 V and 2·2 W.

Calculate the value of resistor R. **5**

Total marks 13

MARKS DO NOT WRITE IN THIS MARGIN

5. A golfer hits a ball from point P. The ball leaves the club with a velocity v at an angle of θ to the horizontal.

The ball travels through the air and lands at point R.

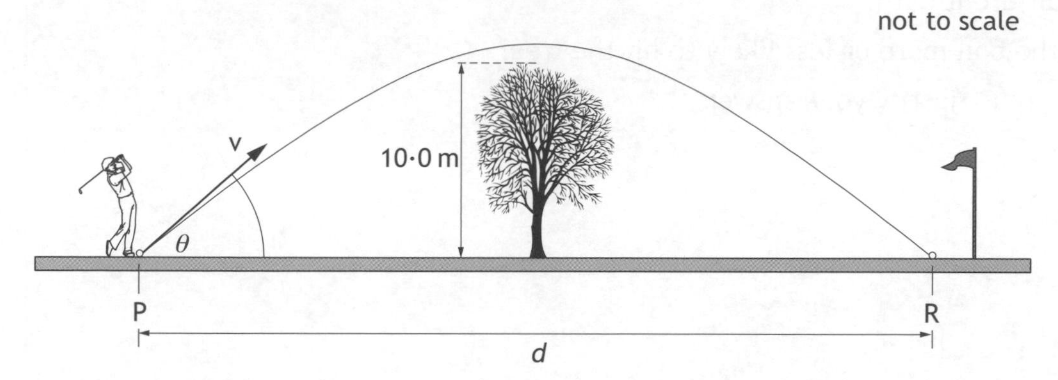

Midway between P and R there is a tree of height 10·0 m.

(a) The horizontal and vertical components of the ball's velocity during its flight are shown.

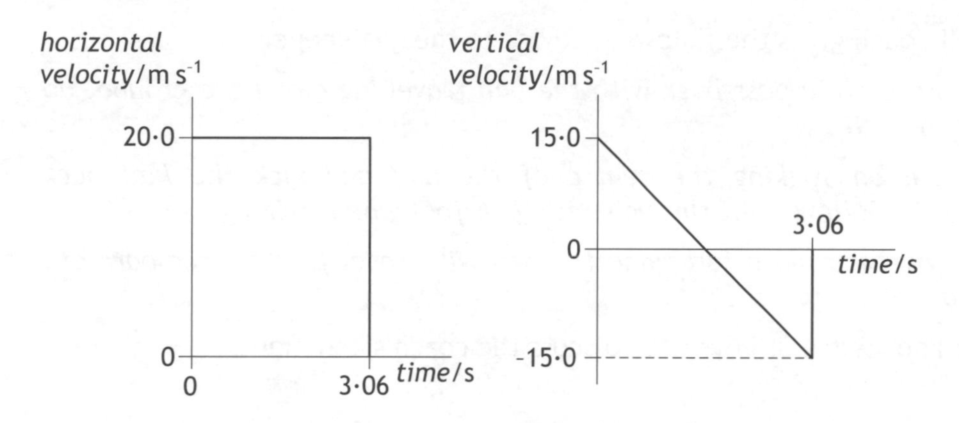

The effects of air resistance can be ignored.

Calculate:

(i) the horizontal distance *d*; **2**

(ii) the maximum height of the ball above the ground. **3**

MARKS DO NOT WRITE IN THIS MARGIN

5. (continued)

(b) When the effects of air resistance are not ignored, the golf ball follows a different path.

Is the ball more or less likely to hit the tree? **2**

You must justify your answer.

Total marks 7

6. A football coach gives the following advice to the goalkeeper.

When you receive a pass back with the ball travelling along the ground, do not stop the ball.

Concentrate on striking the centre of the ball and kick the ball back immediately. Follow right through with your foot when striking the ball.

This will give you more force and the ball will travel further, compared to striking a stationary ball.

Use your knowledge of Physics to discuss the coach's comments. **3**

MARKS | DO NOT WRITE IN THIS MARGIN

7. Ultraviolet radiation from a lamp is incident on the surface of a metal.

This causes the release of electrons from the surface of the metal.

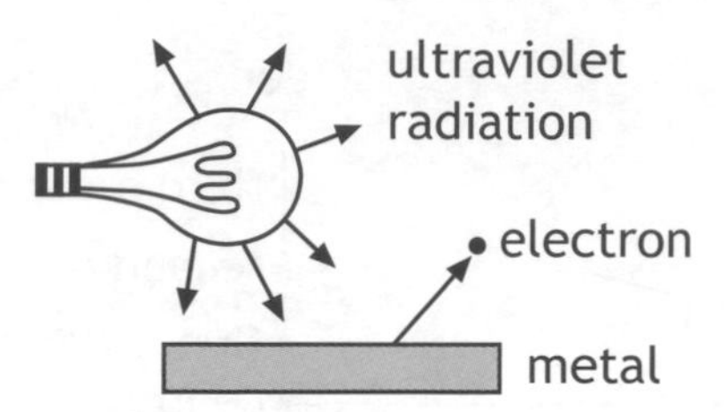

The energy of each photon of ultraviolet light is $5{\cdot}23 \times 10^{-19}$ J.

The work function of the metal is $2{\cdot}56 \times 10^{-19}$ J.

(a) Calculate:

(i) the maximum kinetic energy of an electron released from this metal by this radiation; **1**

(ii) the maximum speed of an emitted electron. **3**

(b) The source of ultraviolet radiation is now moved further away from the surface of the metal.

State the effect, if any, this has on the maximum speed of an emitted electron. **2**

Justify your answer.

Total marks 6

MARKS | DO NOT WRITE IN THIS MARGIN

8. A manufacturer claims that a grating consists of $3 \cdot 00 \times 10^5$ lines per metre and is accurate to ± 2·0%. A technician decides to test this claim. She directs laser light of wavelength 633 nm onto the grating.

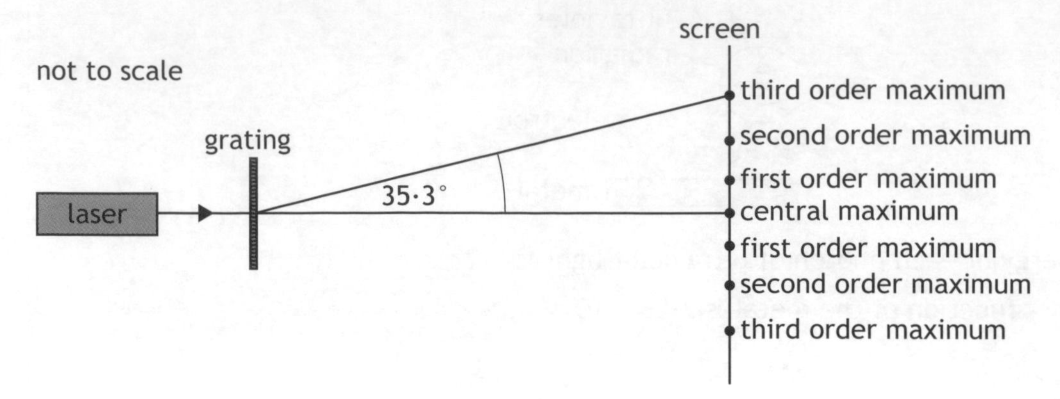

She measures the angle between the central maximum and the third order maximum to be 35·3°.

(a) Calculate the value she obtains for the slit separation for this grating. **3**

(b) What value does she determine for the number of lines per metre for this grating? **2**

MARKS DO NOT WRITE IN THIS MARGIN

8. (continued)

(c) Does the technician's value for the number of lines per metre agree with the manufacturer's claim of $3{\cdot}00 \times 10^5$ lines per metre ± 2·0%? **3**

You must justify your answer by calculation.

Total marks 8

9. (a) Electrons which orbit the nucleus of an atom can be considered as occupying discrete energy levels.

The following diagram shows some of the energy levels for a particular atom.

E_3 ———————— $-5{\cdot}2 \times 10^{-19}$ J

E_2 ———————— $-9{\cdot}0 \times 10^{-19}$ J

E_1 ———————— $-16{\cdot}2 \times 10^{-19}$ J

E_0 ———————— $-24{\cdot}6 \times 10^{-19}$ J

(i) Radiation is produced when electrons make transitions from a higher to a lower energy level.

Which transition, between these energy levels, produces radiation with the shortest wavelength? **2**

Justify your answer.

MARKS | DO NOT WRITE IN THIS MARGIN

9. (a) (continued)

(ii) An electron is excited from energy level E_2 to E_3 by absorbing light energy.

What frequency of light is used to excite this electron? **4**

(b) Another source of light has a frequency of $4{\cdot}6 \times 10^{14}$ Hz in air.

A ray of this light is directed into a block of transparent material as shown.

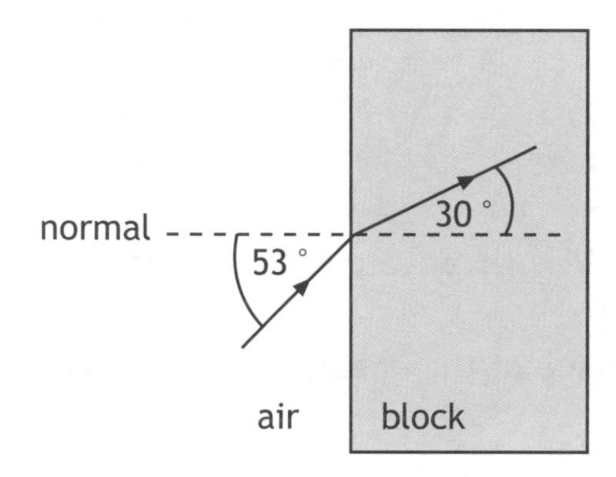

Calculate the wavelength of the light in the block. **6**

Total marks 12

MARKS | DO NOT WRITE IN THIS MARGIN

10. The apparatus shown in the diagram is designed to accelerate alpha particles.

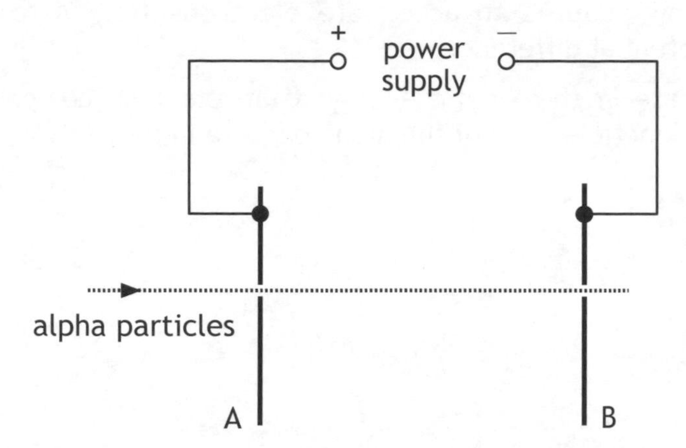

An alpha particle travelling at a speed of $2{\cdot}60 \times 10^{6}$ ms^{-1} passes through a hole in plate A. The mass of an alpha particle is $6{\cdot}64 \times 10^{-27}$ kg and its charge is 3.2×10^{-19} C.

(a) When the alpha particle reaches plate B, its kinetic energy has increased to 3.05×10^{-14} J.

Show that the work done on the alpha particle as it moves from plate A to plate B is 8.1×10^{-15} J. **3**

(b) Calculate the potential difference between plates A and B. **3**

MARKS | DO NOT WRITE IN THIS MARGIN

10. (continued)

(c) The apparatus is now adapted to accelerate **electrons** from A to B through the same potential difference.

How does the increase in the kinetic energy of an electron compare with the increase in kinetic energy of the alpha particle in part (a)? **2**

Justify your answer.

Total marks 8

11. A student investigates the relationship between the force exerted on a wire in a magnetic field and the current in the wire.

A pair of magnets is fixed to a yoke and placed on a top pan Newton balance. A rigid copper wire is suspended between the poles of the magnets. The wire is fixed at 90° to the magnetic field.

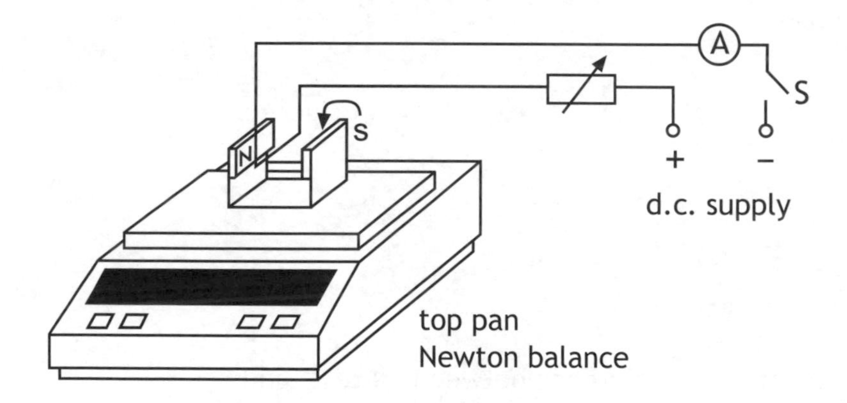

With switch S open the balance is set to zero.

Switch S is closed. The resistor is adjusted and the force recorded for several values of current.

MARKS DO NOT WRITE IN THIS MARGIN

11. (continued)

A graph of force against current is plotted and shown below.

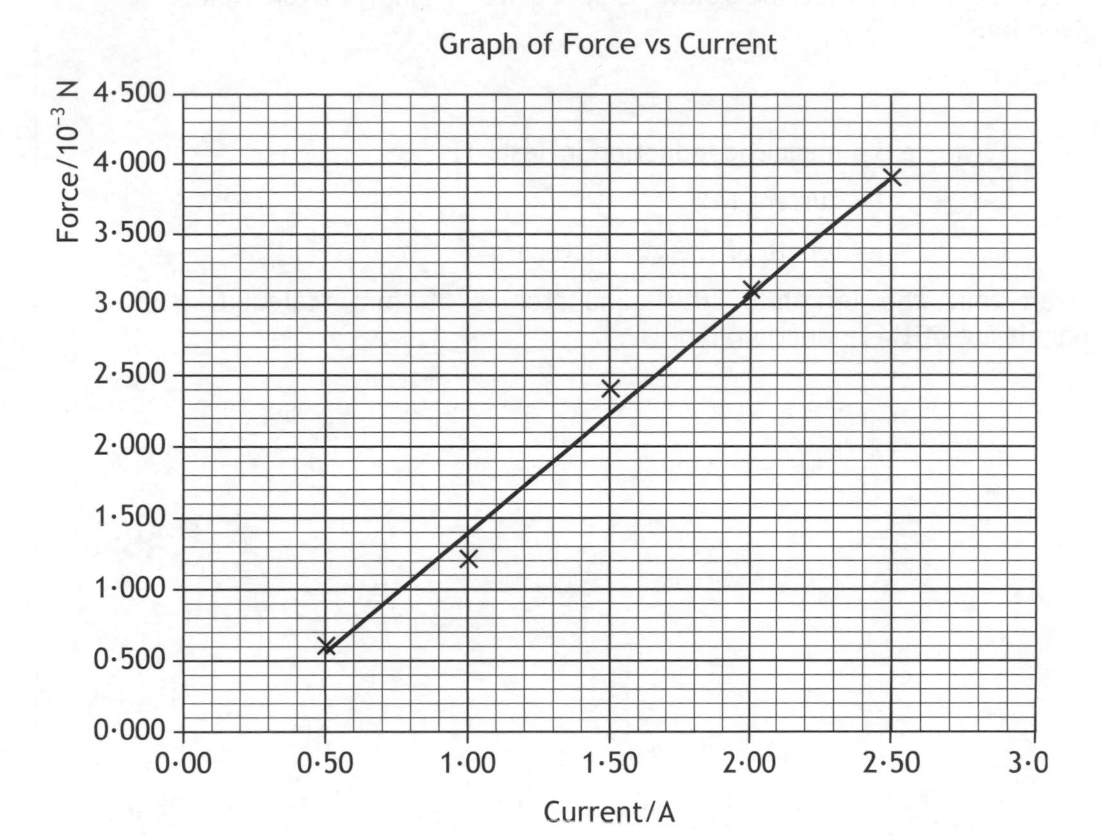

(a) State the direction of the force, due to the magnetic field, acting on the wire. **1**

MARKS DO NOT WRITE IN THIS MARGIN

11. (continued)

(b) It can be shown that the relationship between F in newtons and l is given by:

$$F = B\,I\,l$$

where B = magnetic induction in Tesla (T)

I = current (A)

l = length of the conductor (m)

Given that the length of the conductor is 55 mm, calculate the magnitude of the induction, B. **5**

(c) (i) In the student's evaluation it is stated that the line does not pass through the origin.

Suggest a possible reason for this. **1**

(ii) Suggest one improvement to the experimental procedure that would improve the uncertainty in the results. **1**

Total marks 8

MARKS | DO NOT WRITE IN THIS MARGIN

12. The earth's magnetic field is very important in relation to protection from the solar wind.

The solar wind is the flow of subatomic charged particles from the sun.

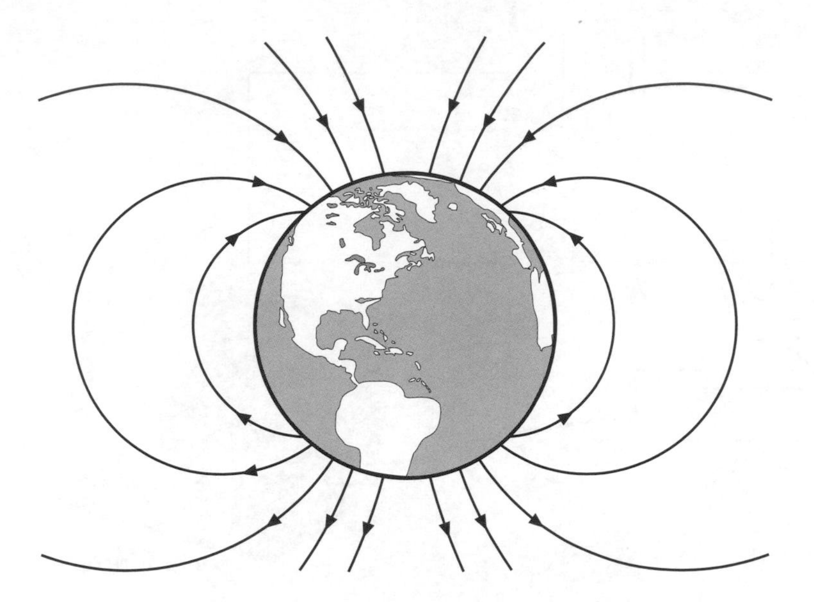

Use your knowledge of Physics to explain this magnetic effect on the solar wind. **3**

MARKS | DO NOT WRITE IN THIS MARGIN

13. (a) State what is meant by the term *capacitance*. **1**

(b) An uncharged capacitor, C, is connected in a circuit as shown.

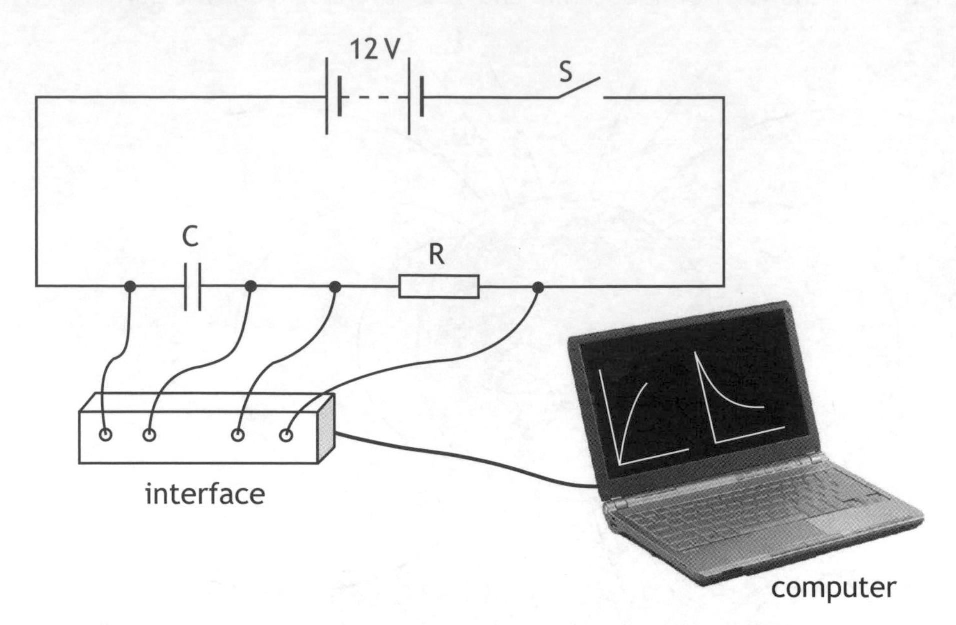

The 12 V battery has negligible internal resistance.

Switch S is closed and the capacitor begins to charge.

The interface measures the current in the circuit and the potential difference (p.d.) across the capacitor. These measurements are displayed as graphs on the computer.

Graph 1 shows the p.d. across the capacitor for the first 0·40 s of charging.

Graph 2 shows the current in the circuit for the first 0·40 s of charging.

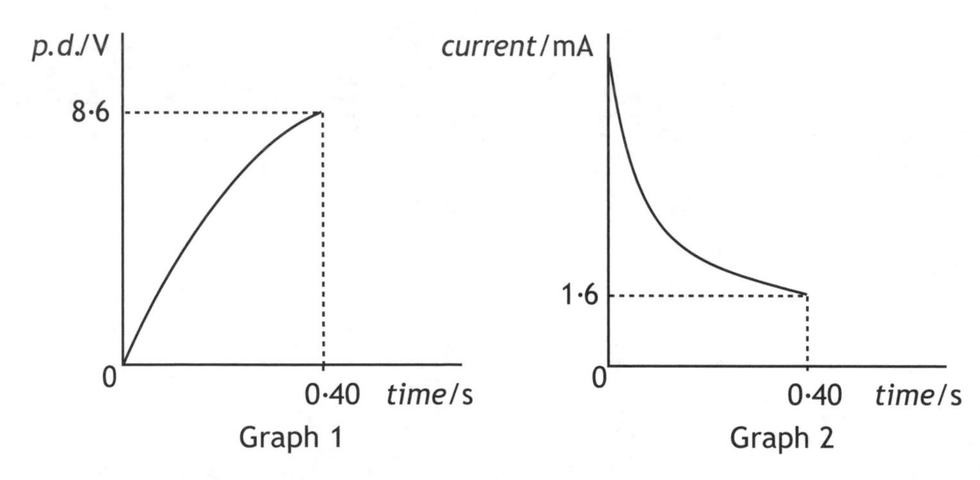

(i) Determine the p.d. **across resistor R** at 0·40 s. **1**

(ii) Calculate the resistance of R. **3**

MARKS DO NOT WRITE IN THIS MARGIN

13. (b) (continued)

(iii) The capacitor takes 2·2 seconds to charge fully.

At that time it stores 10·8 mJ of energy.

Calculate the capacitance of the capacitor. **4**

(c) The capacitor is now discharged.

A second, identical resistor is connected in the circuit as shown.

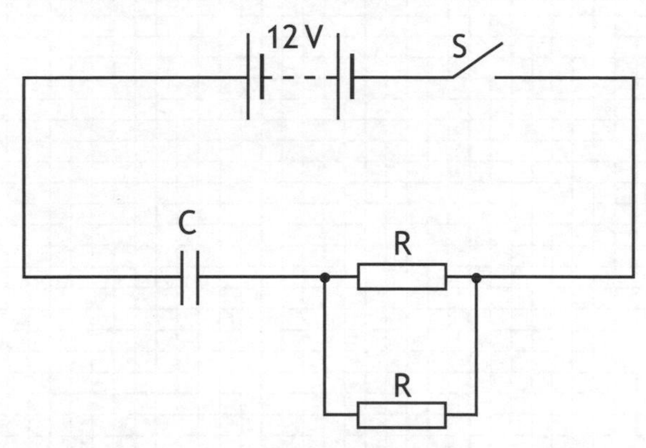

Switch S is closed.

Is the time taken for the capacitor to fully charge less than, equal to, or greater than the time taken to fully charge in part (b)? **2**

Justify your answer.

Total marks 11

[END OF MODEL PAPER]

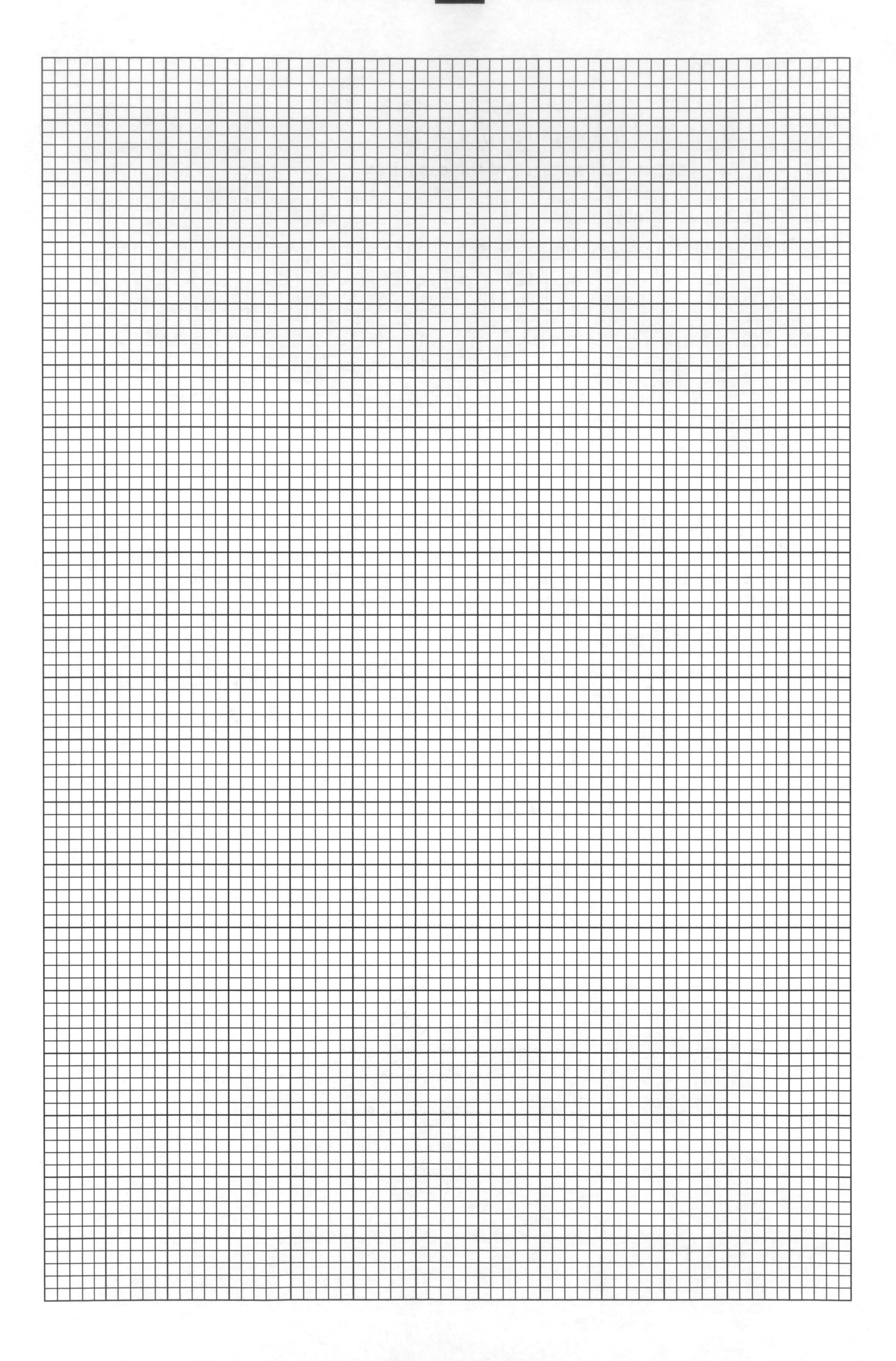

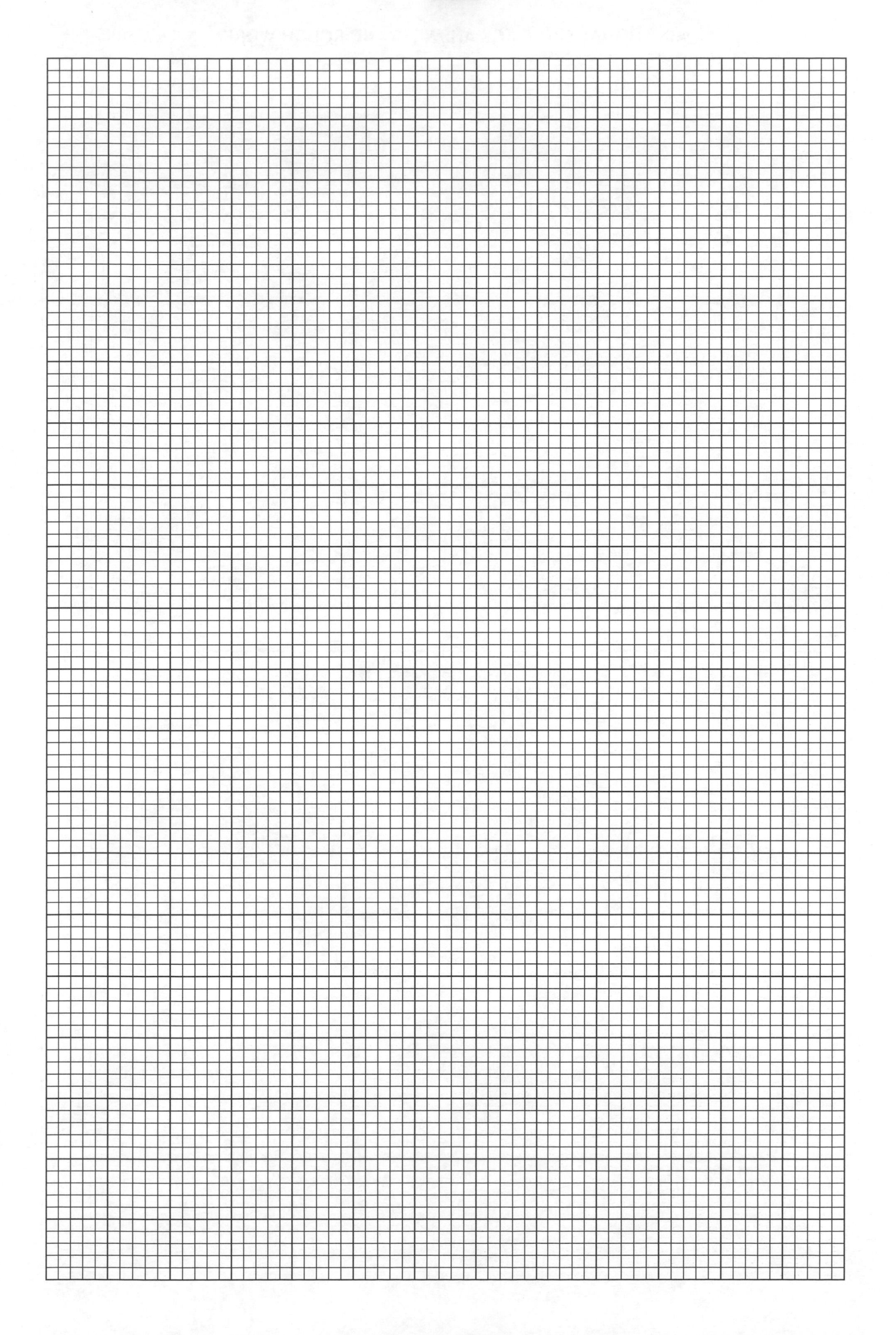

ADDITIONAL SPACE FOR ANSWERS AND ROUGH WORK

MARKS | DO NOT WRITE IN THIS MARGIN

ADDITIONAL SPACE FOR ANSWERS AND ROUGH WORK

MARKS | DO NOT WRITE IN THIS MARGIN

ADDITIONAL SPACE FOR ANSWERS AND ROUGH WORK

MARKS | DO NOT WRITE IN THIS MARGIN

HIGHER FOR CfE

2015

X757/76/02

Physics
Section 1—Questions

TUESDAY, 5 MAY

1:00 PM – 3:30 PM

Instructions for the completion of Section 1 are given on *Page two* of your question and answer booklet X757/76/01.

Record your answers on the answer grid on *Page three* of your question and answer booklet.

Reference may be made to the Data Sheet on *Page two* of this booklet and to the Relationships Sheet X757/76/11.

Before leaving the examination room you must give your question and answer booklet to the Invigilator; if you do not, you may lose all the marks for this paper.

DATA SHEET

COMMON PHYSICAL QUANTITIES

Quantity	*Symbol*	*Value*	*Quantity*	*Symbol*	*Value*
Speed of light in vacuum	c	$3·00 \times 10^{8}$ m s^{-1}	Planck's constant	h	$6·63 \times 10^{-34}$ J s
Magnitude of the charge on an electron	e	$1·60 \times 10^{-19}$ C	Mass of electron	m_e	$9·11 \times 10^{-31}$ kg
Universal Constant of Gravitation	G	$6·67 \times 10^{-11}$ m^{3} kg^{-1} s^{-2}	Mass of neutron	m_n	$1·675 \times 10^{-27}$ kg
Gravitational acceleration on Earth	g	$9·8$ m s^{-2}	Mass of proton	m_p	$1·673 \times 10^{-27}$ kg
Hubble's constant	H_0	$2·3 \times 10^{-18}$ s^{-1}			

REFRACTIVE INDICES

The refractive indices refer to sodium light of wavelength 589 nm and to substances at a temperature of 273 K.

Substance	*Refractive index*	*Substance*	*Refractive index*
Diamond	2·42	Water	1·33
Crown glass	1·50	Air	1·00

SPECTRAL LINES

Element	*Wavelength*/nm	*Colour*
Hydrogen	656	Red
	486	Blue-green
	434	Blue-violet
	410	Violet
	397	Ultraviolet
	389	Ultraviolet
Sodium	589	Yellow

Element	*Wavelength*/nm	*Colour*
Cadmium	644	Red
	509	Green
	480	Blue

Lasers

Element	*Wavelength*/nm	*Colour*
Carbon dioxide	9550 } 10590 }	Infrared
Helium-neon	633	Red

PROPERTIES OF SELECTED MATERIALS

Substance	*Density*/kg m^{-3}	*Melting Point*/K	*Boiling Point*/K
Aluminium	$2·70 \times 10^{3}$	933	2623
Copper	$8·96 \times 10^{3}$	1357	2853
Ice	$9·20 \times 10^{2}$	273	. . . .
Sea Water	$1·02 \times 10^{3}$	264	377
Water	$1·00 \times 10^{3}$	273	373
Air	1·29	. . . .	. . . .
Hydrogen	$9·0 \times 10^{-2}$	14	20

The gas densities refer to a temperature of 273 K and a pressure of $1·01 \times 10^{5}$ Pa.

SECTION 1 — 20 marks

Attempt ALL questions

1. The following velocity-time graph represents the vertical motion of a ball.

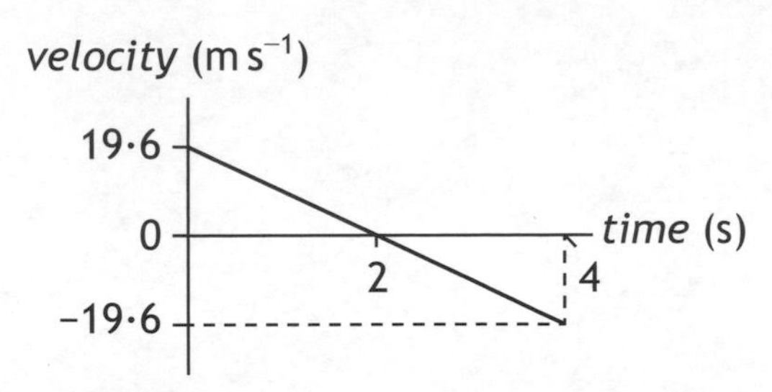

Which of the following acceleration-time graphs represents the same motion?

A

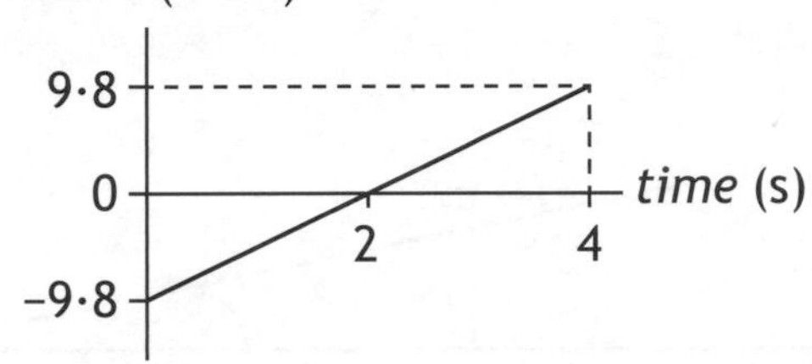

B

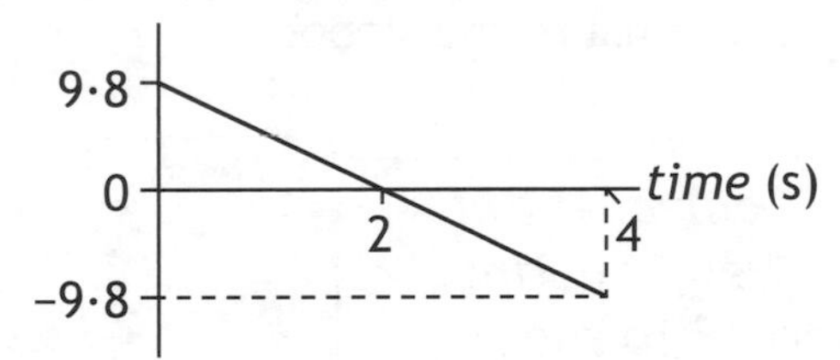

C

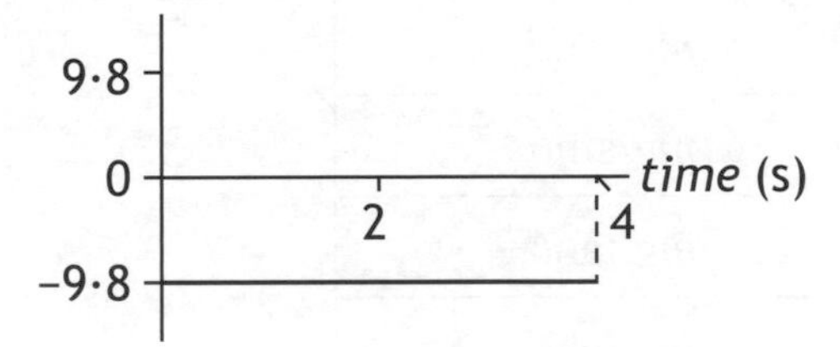

D

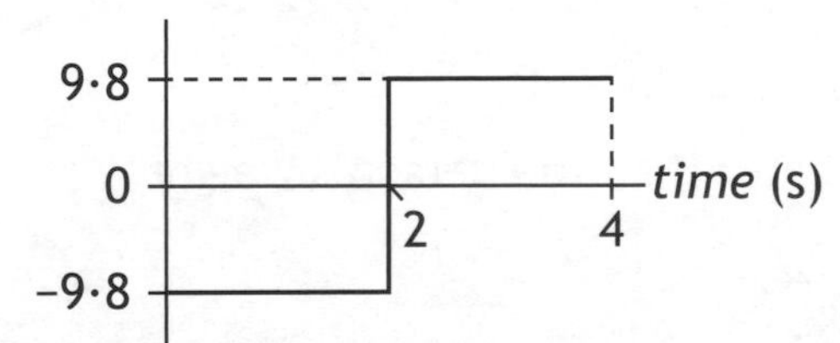

E

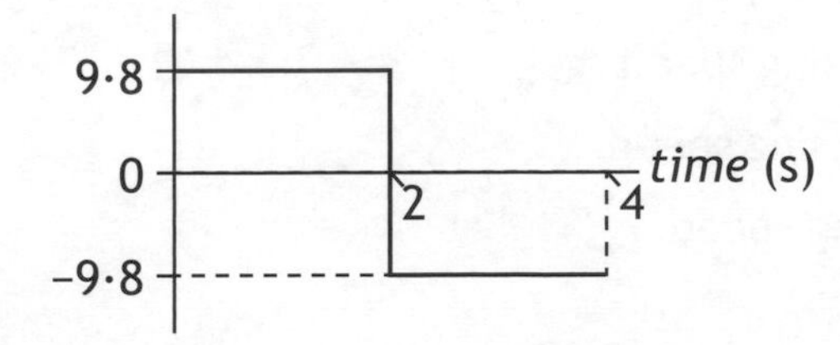

2. A car is travelling at $12\,m\,s^{-1}$ along a straight road. The car now accelerates uniformly at $-1{\cdot}5\,m\,s^{-2}$ for $6{\cdot}0\,s$.

The distance travelled during this time is

A 18 m

B 45 m

C 68 m

D 72 m

E 99 m.

3. A box of mass m rests on a slope as shown.

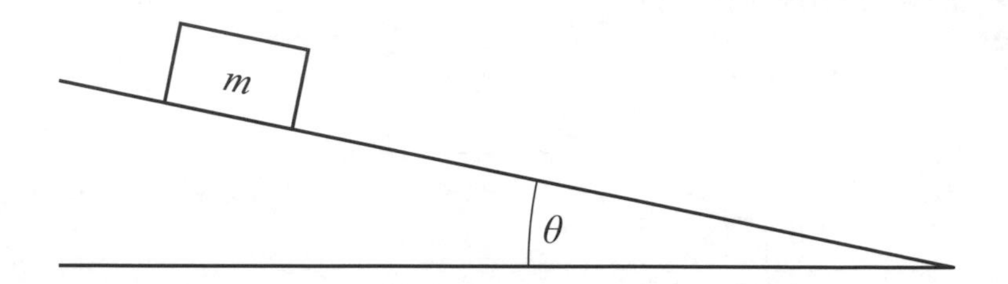

Which row in the table shows the component of the weight acting down the slope and the component of the weight acting normal to the slope?

	Component of weight acting down the slope	*Component of weight acting normal to the slope*
A	$mg\sin\theta$	$mg\cos\theta$
B	$mg\tan\theta$	$mg\sin\theta$
C	$mg\cos\theta$	$mg\sin\theta$
D	$mg\cos\theta$	$mg\tan\theta$
E	$mg\sin\theta$	$mg\tan\theta$

4. A person stands on bathroom scales in a lift.

The scales show a reading greater than the person's weight.

The lift is moving

A upwards with constant speed

B downwards with constant speed

C downwards with increasing speed

D downwards with decreasing speed

E upwards with decreasing speed.

5. A car of mass 900 kg pulls a caravan of mass 400 kg along a straight, horizontal road with an acceleration of $2{\cdot}0\,m\,s^{-2}$.

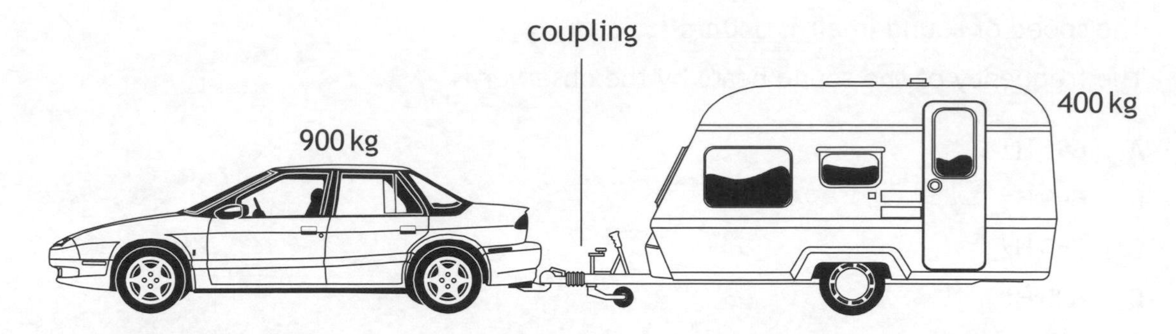

Assuming that the frictional forces on the caravan are negligible, the tension in the coupling between the car and the caravan is

A 400 N

B 500 N

C 800 N

D 1800 N

E 2600 N.

6. Water flows at a rate of $6{\cdot}25 \times 10^8$ kg per minute over a waterfall.

The height of the waterfall is 108 m.

The total power delivered by the water in falling through the 108 m is

A $1{\cdot}13 \times 10^9$ W

B $1{\cdot}10 \times 10^{10}$ W

C $6{\cdot}62 \times 10^{11}$ W

D $4{\cdot}05 \times 10^{12}$ W

E $3{\cdot}97 \times 10^{13}$ W.

7. A spacecraft is travelling at a constant speed of $0{\cdot}60c$ relative to the Moon.

An observer on the Moon measures the length of the moving spacecraft to be 190 m.

The length of the spacecraft as measured by an astronaut on the spacecraft is

A 120 m

B 152 m

C 238 m

D 297 m

E 300 m.

[Turn over

8. A siren on an ambulance emits sound at a constant frequency of 750 Hz.

The ambulance is travelling at a constant speed of 25·0 m s^{-1} towards a stationary observer.

The speed of sound in air is 340 m s^{-1}.

The frequency of the sound heard by the observer is

A 695 Hz

B 699 Hz

C 750 Hz

D 805 Hz

E 810 Hz.

9. The emission of beta particles in radioactive decay is evidence for the existence of

A quarks

B electrons

C gluons

D neutrinos

E bosons.

10. Two parallel metal plates X and Y in a vacuum have a potential difference V across them.

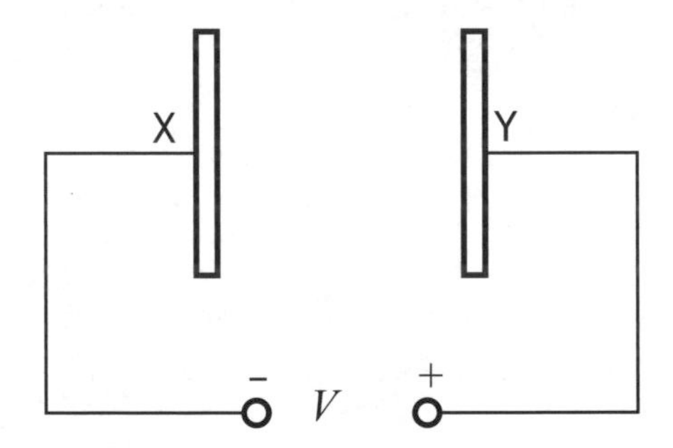

An electron of charge e and mass m, initially at rest, is released from plate X.

The speed of the electron when it reaches plate Y is given by

A $\frac{2eV}{m}$

B $\sqrt{\frac{2eV}{m}}$

C $\sqrt{\frac{2V}{em}}$

D $\frac{2V}{em}$

E $\frac{2mV}{e}$

11. A potential difference of 2 kV is applied across two metal plates.

An electron passes between the metal plates and follows the path shown.

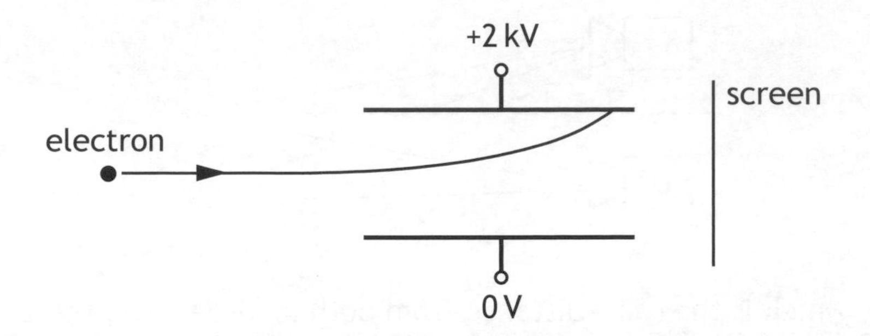

A student makes the following statements about changes that could be made to allow the electron to pass between the plates and reach the screen.

I Increasing the initial speed of the electron could allow the electron to reach the screen.

II Increasing the potential difference across the plates could allow the electron to reach the screen.

III Reversing the polarity of the plates could allow the electron to reach the screen.

Which of these statements is/are correct?

A I only

B II only

C III only

D I and II only

E I and III only

12. The following statement describes a fusion reaction.

$${}^{2}_{1}H + {}^{2}_{1}H \rightarrow {}^{3}_{2}He + {}^{1}_{0}n + \text{energy}$$

The total mass of the particles before the reaction is $6{\cdot}684 \times 10^{-27}$ kg.

The total mass of the particles after the reaction is $6{\cdot}680 \times 10^{-27}$ kg.

The energy released in the reaction is

A $6{\cdot}012 \times 10^{-10}$ J

B $6{\cdot}016 \times 10^{-10}$ J

C $1{\cdot}800 \times 10^{-13}$ J

D $3{\cdot}600 \times 10^{-13}$ J

E $1{\cdot}200 \times 10^{-21}$ J.

[Turn over

13. Two identical loudspeakers, L_1 and L_2, are operated at the same frequency and in phase with each other. An interference pattern is produced.

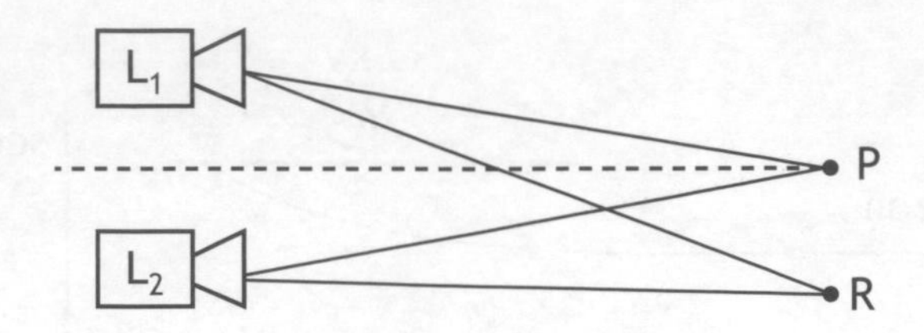

At position P, which is the same distance from both loudspeakers, there is a maximum.

The next maximum is at position R, where $L_1R = 5{\cdot}6\,m$ and $L_2R = 5{\cdot}3\,m$.

The speed of sound in air is $340\,m\,s^{-1}$.

The frequency of the sound emitted by the loudspeakers is

A $8{\cdot}8 \times 10^{-4}\,Hz$

B $3{\cdot}1 \times 10^{1}\,Hz$

C $1{\cdot}0 \times 10^{2}\,Hz$

D $1{\cdot}1 \times 10^{3}\,Hz$

E $3{\cdot}7 \times 10^{3}\,Hz$.

14. An experiment is carried out to measure the wavelength of red light from a laser.

The following values for the wavelength are obtained.

650 nm 640 nm 635 nm 648 nm 655 nm

The mean value for the wavelength and the approximate random uncertainty in the mean is

A (645 ± 1) nm

B (645 ± 4) nm

C (646 ± 1) nm

D (646 ± 4) nm

E (3228 ± 20) nm.

15. Red light is used to investigate the critical angle of two materials P and Q.

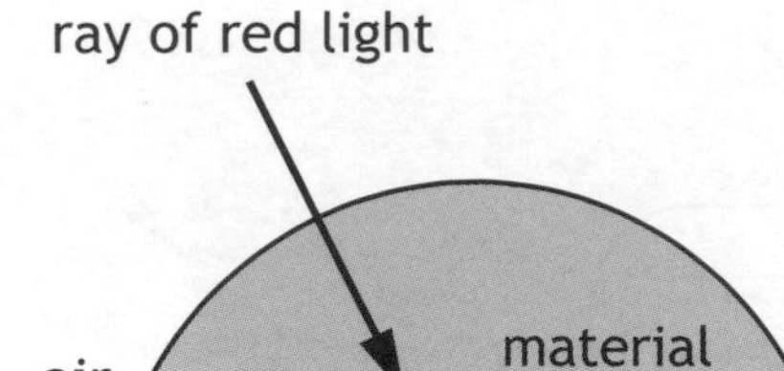

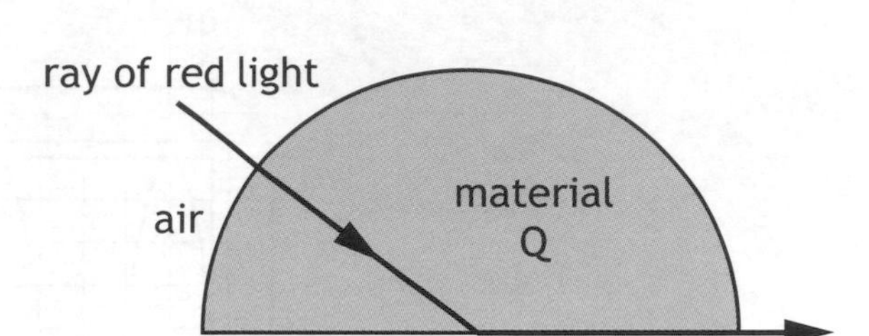

A student makes the following statements.

I Material P has a higher refractive index than material Q.

II The wavelength of the red light is longer inside material P than inside material Q.

III The red light travels at the same speed inside materials P and Q.

Which of these statements is/are correct?

A I only

B II only

C III only

D I and II only

E I, II and III

16. The diagram represents some electron transitions between energy levels in an atom.

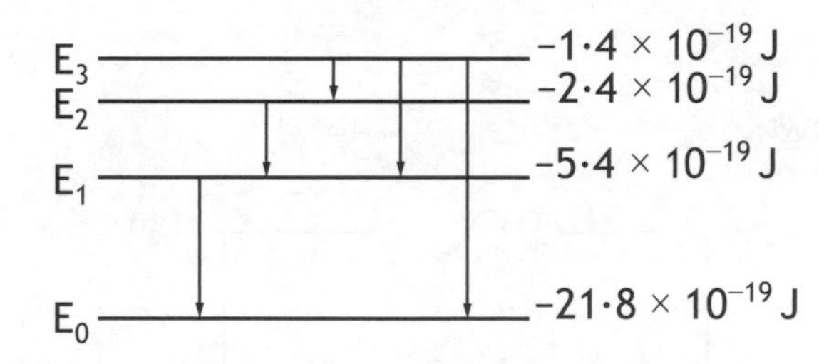

The radiation emitted with the shortest wavelength is produced by an electron making transition

A E_1 to E_0

B E_2 to E_1

C E_3 to E_2

D E_3 to E_1

E E_3 to E_0.

[Turn over

17. The output from a signal generator is connected to the input terminals of an oscilloscope. The trace observed on the oscilloscope screen, the Y-gain setting and the timebase setting are shown.

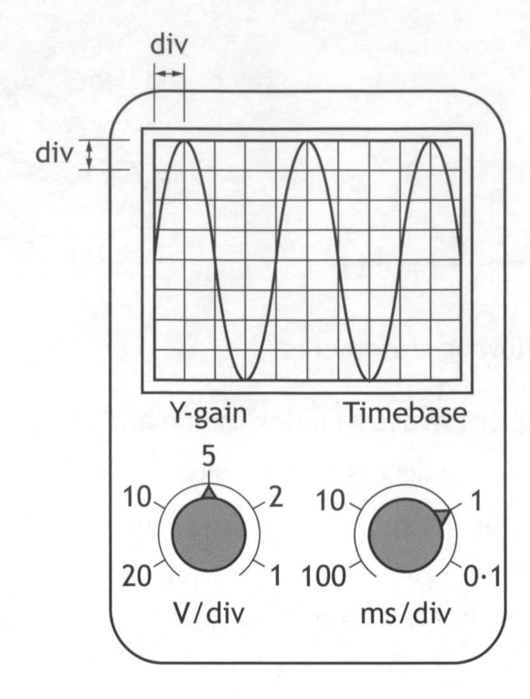

The frequency of the signal shown is calculated using the

A timebase setting and the vertical height of the trace

B timebase setting and the horizontal distance between the peaks of the trace

C Y-gain setting and the vertical height of the trace

D Y-gain setting and the horizontal distance between the peaks of the trace

E Y-gain setting and the timebase setting.

18. A circuit is set up as shown.

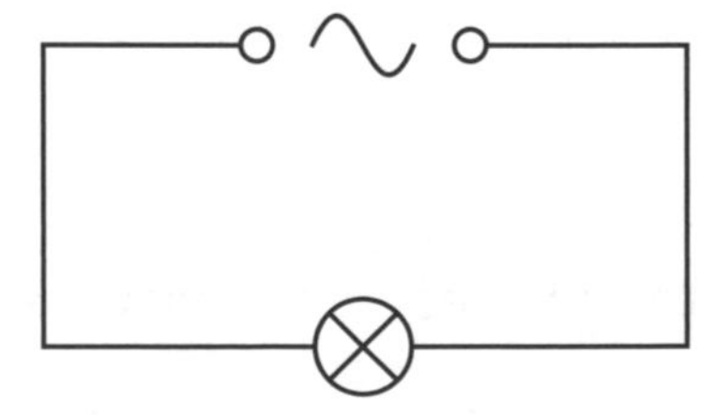

The r.m.s voltage across the lamp is 12 V.

The power produced by the lamp is 24 W.

The peak current in the lamp is

A 0·71 A

B 1·4 A

C 2·0 A

D 2·8 A

E 17 A.

19. A student makes the following statements about energy bands in different materials.

I In metals the highest occupied energy band is not completely full.

II In insulators the highest occupied energy band is full.

III The gap between the valence band and conduction band is smaller in semiconductors than in insulators.

Which of these statements is/are correct?

A I only

B II only

C I and II only

D I and III only

E I, II and III

20. The upward lift force L on the wings of an aircraft is calculated using the relationship

$$L = \tfrac{1}{2}\rho v^2 A C_L$$

where:

ρ is the density of air
v is the speed of the wings through the air
A is the area of the wings
C_L is the coefficient of lift.

The weight of a model aircraft is 80·0 N.
The area of the wings on the model aircraft is 3·0 m^2.
The coefficient of lift for these wings is 1·6.
The density of air is 1·29 kg m^{-3}

The speed required for the model aircraft to maintain a level flight is

A 2·5 m s^{-1}

B 3·6 m s^{-1}

C 5·1 m s^{-1}

D 12·9 m s^{-1}

E 25·8 m s^{-1}.

[END OF SECTION 1. NOW ATTEMPT THE QUESTIONS IN SECTION 2 OF YOUR QUESTION AND ANSWER BOOKLET]

[BLANK PAGE]

DO NOT WRITE ON THIS PAGE

National
Qualifications
2015

X757/76/11

Physics
Relationship Sheet

TUESDAY, 5 MAY
1:00 PM – 3:30 PM

Relationships required for Physics Higher

$d = \bar{v}t$

$s = \bar{v}t$

$v = u + at$

$s = ut + \frac{1}{2}at^2$

$v^2 = u^2 + 2as$

$s = \frac{1}{2}(u+v)t$

$W = mg$

$F = ma$

$E_W = Fd$

$E_p = mgh$

$E_k = \frac{1}{2}mv^2$

$P = \frac{E}{t}$

$p = mv$

$Ft = mv - mu$

$F = G\frac{m_1 m_2}{r^2}$

$t' = \frac{t}{\sqrt{1-\left(\frac{v}{c}\right)^2}}$

$l' = l\sqrt{1-\left(\frac{v}{c}\right)^2}$

$f_o = f_s\left(\frac{v}{v \pm v_s}\right)$

$z = \frac{\lambda_{observed} - \lambda_{rest}}{\lambda_{rest}}$

$z = \frac{v}{c}$

$v = H_0 d$

$W = QV$

$E = mc^2$

$E = hf$

$E_k = hf - hf_0$

$E_2 - E_1 = hf$

$T = \frac{1}{f}$

$v = f\lambda$

$d\sin\theta = m\lambda$

$n = \frac{\sin\theta_1}{\sin\theta_2}$

$\frac{\sin\theta_1}{\sin\theta_2} = \frac{\lambda_1}{\lambda_2} = \frac{v_1}{v_2}$

$\sin\theta_c = \frac{1}{n}$

$I = \frac{k}{d^2}$

$I = \frac{P}{A}$

$V_{peak} = \sqrt{2}V_{rms}$

$I_{peak} = \sqrt{2}I_{rms}$

$Q = It$

$V = IR$

$P = IV = I^2R = \frac{V^2}{R}$

$R_T = R_1 + R_2 + \ldots$

$\frac{1}{R_T} = \frac{1}{R_1} + \frac{1}{R_2} + \ldots$

$E = V + Ir$

$V_1 = \left(\frac{R_1}{R_1 + R_2}\right)V_s$

$\frac{V_1}{V_2} = \frac{R_1}{R_2}$

$C = \frac{Q}{V}$

$E = \frac{1}{2}QV = \frac{1}{2}CV^2 = \frac{1}{2}\frac{Q^2}{C}$

path difference $= m\lambda$ or $\left(m + \frac{1}{2}\right)\lambda$ where $m = 0, 1, 2 \ldots$

random uncertainty $= \frac{\text{max. value} - \text{min. value}}{\text{number of values}}$

Additional Relationships

Circle

circumference $= 2\pi r$

area $= \pi r^2$

Sphere

area $= 4\pi r^2$

volume $= \frac{4}{3}\pi r^3$

Trigonometry

$\sin\theta = \frac{\text{opposite}}{\text{hypotenuse}}$

$\cos\theta = \frac{\text{adjacent}}{\text{hypotenuse}}$

$\tan\theta = \frac{\text{opposite}}{\text{adjacent}}$

$\sin^2\theta + \cos^2\theta = 1$

Electron Arrangements of Elements

Key

Atomic number Symbol Electron arrangement Name

Transition Elements

Group 1 (1)	Group 2 (2)	(3)	(4)	(5)	(6)	(7)	(8)	(9)	(10)	(11)	(12)	Group 3 (13)	Group 4 (14)	Group 5 (15)	Group 6 (16)	Group 7 (17)	Group 0 (18)
1 **H** 1 Hydrogen																	2 **He** 2 Helium
3 **Li** 2,1 Lithium	4 **Be** 2,2 Beryllium											5 **B** 2,3 Boron	6 **C** 2,4 Carbon	7 **N** 2,5 Nitrogen	8 **O** 2,6 Oxygen	9 **F** 2,7 Fluorine	10 **Ne** 2,8 Neon
11 **Na** 2,8,1 Sodium	12 **Mg** 2,8,2 Magnesium											13 **Al** 2,8,3 Aluminium	14 **Si** 2,8,4 Silicon	15 **P** 2,8,5 Phosphorus	16 **S** 2,8,6 Sulfur	17 **Cl** 2,8,7 Chlorine	18 **Ar** 2,8,8 Argon
19 **K** 2,8,8,1 Potassium	20 **Ca** 2,8,8,2 Calcium	21 **Sc** 2,8,9,2 Scandium	22 **Ti** 2,8,10,2 Titanium	23 **V** 2,8,11,2 Vanadium	24 **Cr** 2,8,13,1 Chromium	25 **Mn** 2,8,13,2 Manganese	26 **Fe** 2,8,14,2 Iron	27 **Co** 2,8,15,2 Cobalt	28 **Ni** 2,8,16,2 Nickel	29 **Cu** 2,8,18,1 Copper	30 **Zn** 2,8,18,2 Zinc	31 **Ga** 2,8,18,3 Gallium	32 **Ge** 2,8,18,4 Germanium	33 **As** 2,8,18,5 Arsenic	34 **Se** 2,8,18,6 Selenium	35 **Br** 2,8,18,7 Bromine	36 **Kr** 2,8,18,8 Krypton
37 **Rb** 2,8,18,8,1 Rubidium	38 **Sr** 2,8,18,8,2 Strontium	39 **Y** 2,8,18,9,2 Yttrium	40 **Zr** 2,8,18, 10,2 Zirconium	41 **Nb** 2,8,18, 12,1 Niobium	42 **Mo** 2,8,18,13, 1 Molybdenum	43 **Tc** 2,8,18,13, 2 Technetium	44 **Ru** 2,8,18,15, 1 Ruthenium	45 **Rh** 2,8,18,16, 1 Rhodium	46 **Pd** 2,8,18, 18,0 Palladium	47 **Ag** 2,8,18, 18,1 Silver	48 **Cd** 2,8,18, 18,2 Cadmium	49 **In** 2,8,18, 18,3 Indium	50 **Sn** 2,8,18, 18,4 Tin	51 **Sb** 2,8,18, 18,5 Antimony	52 **Te** 2,8,18, 18,6 Tellurium	53 **I** 2,8,18, 18,7 Iodine	54 **Xe** 2,8,18, 18,8 Xenon
55 **Cs** 2,8,18,18, 8,1 Caesium	56 **Ba** 2,8,18,18, 8,2 Barium	57 **La** 2,8,18,18, 9,2 Lanthanum	72 **Hf** 2,8,18,32, 10,2 Hafnium	73 **Ta** 2,8,18, 32,11,2 Tantalum	74 **W** 2,8,18,32, 12,2 Tungsten	75 **Re** 2,8,18,32, 13,2 Rhenium	76 **Os** 2,8,18,32, 14,2 Osmium	77 **Ir** 2,8,18,32, 15,2 Iridium	78 **Pt** 2,8,18,32, 17,1 Platinum	79 **Au** 2,8,18, 32,18,1 Gold	80 **Hg** 2,8,18, 32,18,2 Mercury	81 **Tl** 2,8,18, 32,18,3 Thallium	82 **Pb** 2,8,18, 32,18,4 Lead	83 **Bi** 2,8,18, 32,18,5 Bismuth	84 **Po** 2,8,18, 32,18,6 Polonium	85 **At** 2,8,18, 32,18,7 Astatine	86 **Rn** 2,8,18, 32,18,8 Radon
87 **Fr** 2,8,18,32, 18,8,1 Francium	88 **Ra** 2,8,18,32, 18,8,2 Radium	89 **Ac** 2,8,18,32, 18,9,2 Actinium	104 **Rf** 2,8,18,32, 32,10,2 Rutherfordium	105 **Db** 2,8,18,32, 32,11,2 Dubnium	106 **Sg** 2,8,18,32, 32,12,2 Seaborgium	107 **Bh** 2,8,18,32, 32,13,2 Bohrium	108 **Hs** 2,8,18,32, 32,14,2 Hassium	109 **Mt** 2,8,18,32, 32,15,2 Meitnerium	110 **Ds** 2,8,18,32, 32,17,1 Darmstadtium	111 **Rg** 2,8,18,32, 32,18,1 Roentgenium	112 **Cn** 2,8,18,32, 32,18,2 Copernicium						

Lanthanides	57 **La** 2,8,18, 18,9,2 Lanthanum	58 **Ce** 2,8,18, 20,8,2 Cerium	59 **Pr** 2,8,18,21, 8,2 Praseodymium	60 **Nd** 2,8,18,22, 8,2 Neodymium	61 **Pm** 2,8,18,23, 8,2 Promethium	62 **Sm** 2,8,18,24, 8,2 Samarium	63 **Eu** 2,8,18,25, 8,2 Europium	64 **Gd** 2,8,18,25, 9,2 Gadolinium	65 **Tb** 2,8,18,27, 8,2 Terbium	66 **Dy** 2,8,18,28, 8,2 Dysprosium	67 **Ho** 2,8,18,29, 8,2 Holmium	68 **Er** 2,8,18,30, 8,2 Erbium	69 **Tm** 2,8,18,31, 8,2 Thulium	70 **Yb** 2,8,18,32, 8,2 Ytterbium	71 **Lu** 2,8,18,32, 9,2 Lutetium
Actinides	89 **Ac** 2,8,18,32, 18,9,2 Actinium	90 **Th** 2,8,18,32, 18,10,2 Thorium	91 **Pa** 2,8,18,32, 20,9,2 Protactinium	92 **U** 2,8,18,32, 21,9,2 Uranium	93 **Np** 2,8,18,32, 22,9,2 Neptunium	94 **Pu** 2,8,18,32, 24,8,2 Plutonium	95 **Am** 2,8,18,32, 25,8,2 Americium	96 **Cm** 2,8,18,32, 25,9,2 Curium	97 **Bk** 2,8,18,32, 27,8,2 Berkelium	98 **Cf** 2,8,18,32, 28,8,2 Californium	99 **Es** 2,8,18,32, 29,8,2 Einsteinium	100 **Fm** 2,8,18,32, 30,8,2 Fermium	101 **Md** 2,8,18,32, 31,8,2 Mendelevium	102 **No** 2,8,18,32, 32,8,2 Nobelium	103 **Lr** 2,8,18,32, 32,9,2 Lawrencium

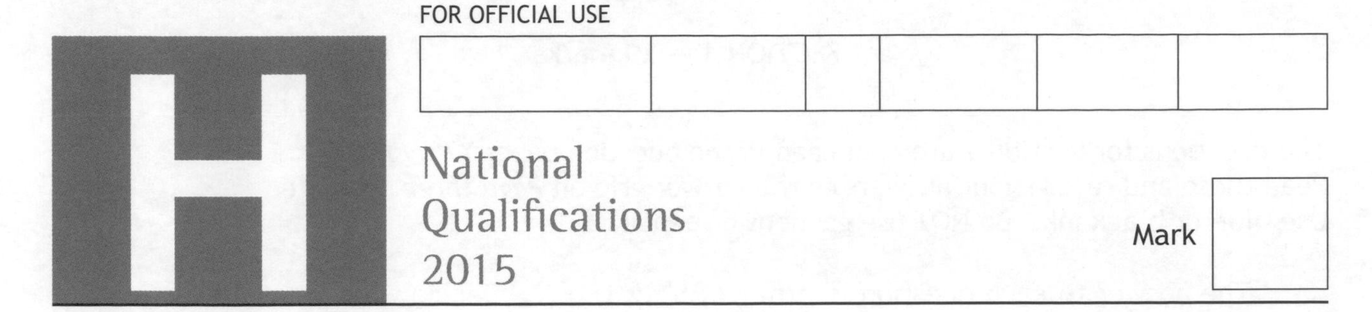

X757/76/01

Physics
Section 1 – Answer Grid and Section 2

TUESDAY, 5 MAY

1:00 PM – 3:30 PM

Fill in these boxes and read what is printed below.

Full name of centre

Town

Forename(s)

Surname

Number of seat

Date of birth

Day Month Year

Scottish candidate number

Total marks — 130

SECTION 1 — 20 marks

Attempt ALL questions.

Instructions for the completion of Section 1 are given on *Page two*.

SECTION 2 — 110 marks

Attempt ALL questions.

Reference may be made to the Data Sheet on *Page two* of the question paper X757/76/02 and to the Relationship Sheet X757/76/11.

Care should be taken to give an appropriate number of significant figures in the final answers to calculations.

Write your answers clearly in the spaces provided in this booklet. Additional space for answers and rough work is provided at the end of this booklet. If you use this space you must clearly identify the question number you are attempting. Any rough work must be written in this booklet. You should score through your rough work when you have written your final copy.

Use **blue** or **black** ink.

Before leaving the examination room you must give this booklet to the Invigilator; if you do not, you may lose all the marks for this paper.

SECTION 1 — 20 marks

The questions for Section 1 are contained in the question paper X757/76/02.
Read these and record your answers on the answer grid on *Page three* opposite.
Use **blue** or **black** ink. Do NOT use gel pens or pencil.

1. The answer to each question is **either** A, B, C, D or E. Decide what your answer is, then fill in the appropriate bubble (see sample question below).

2. There is **only one correct** answer to each question.

3. Any rough work must be written in the additional space for answers and rough work at the end of this booklet.

Sample Question

The energy unit measured by the electricity meter in your home is the:

A ampere

B kilowatt-hour

C watt

D coulomb

E volt.

The correct answer is **B**—kilowatt-hour. The answer **B** bubble has been clearly filled in (see below).

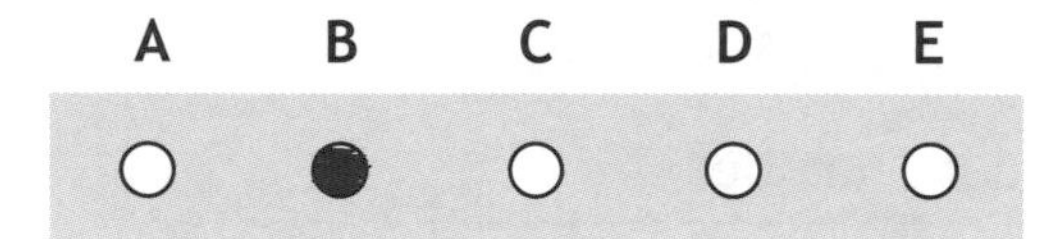

Changing an answer

If you decide to change your answer, cancel your first answer by putting a cross through it (see below) and fill in the answer you want. The answer below has been changed to **D**.

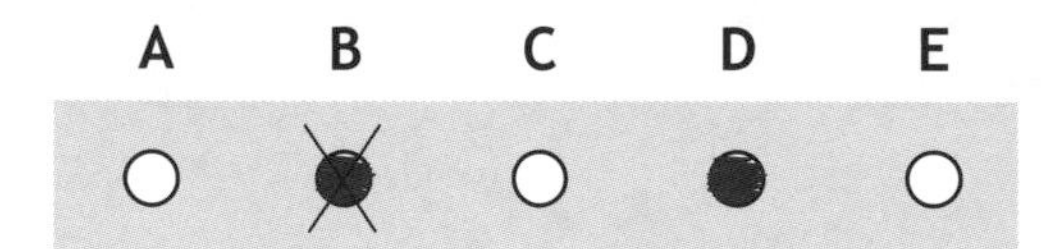

If you then decide to change back to an answer you have already scored out, put a tick (✓) to the **right** of the answer you want, as shown below:

A B C D E

or

A B C D E

SECTION 1 — Answer Grid

	A	B	C	D	E
1	○	○	○	○	○
2	○	○	○	○	○
3	○	○	○	○	○
4	○	○	○	○	○
5	○	○	○	○	○
6	○	○	○	○	○
7	○	○	○	○	○
8	○	○	○	○	○
9	○	○	○	○	○
10	○	○	○	○	○
11	○	○	○	○	○
12	○	○	○	○	○
13	○	○	○	○	○
14	○	○	○	○	○
15	○	○	○	○	○
16	○	○	○	○	○
17	○	○	○	○	○
18	○	○	○	○	○
19	○	○	○	○	○
20	○	○	○	○	○

[BLANK PAGE]

DO NOT WRITE ON THIS PAGE

[Turn over for SECTION 2 on *Page six*

DO NOT WRITE ON THIS PAGE

MARKS | DO NOT WRITE IN THIS MARGIN

SECTION 2 — 110 marks

Attempt ALL questions

1. The shot put is an athletics event in which competitors "throw" a shot as far as possible. The shot is a metal ball of mass 4·0 kg. One of the competitors releases the shot at a height of 1·8 m above the ground and at an angle θ to the horizontal. The shot travels through the air and hits the ground at X. The effects of air resistance are negligible.

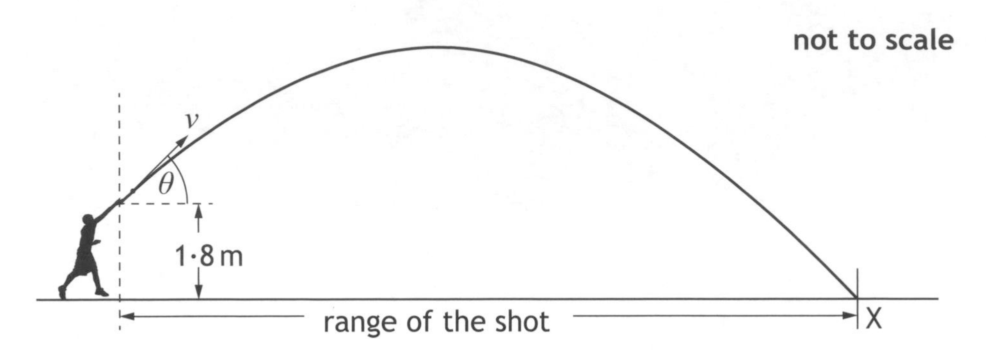

The graph shows how the release speed of the shot v varies with the angle of projection θ.

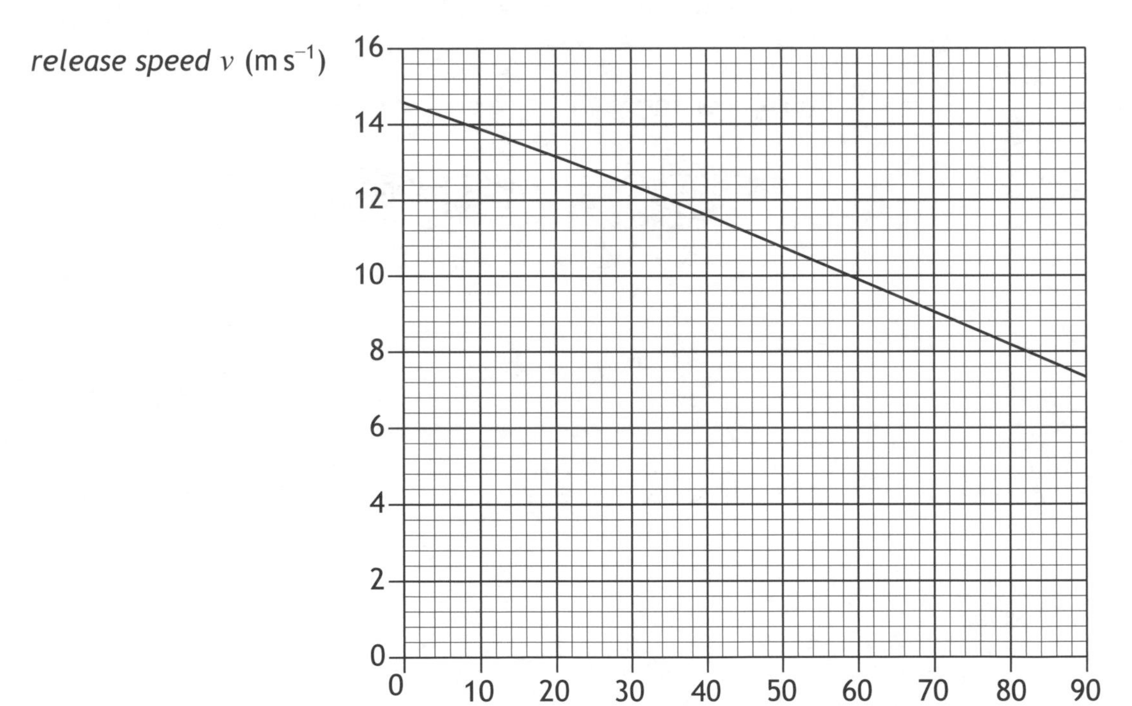

MARKS DO NOT WRITE IN THIS MARGIN

1. (continued)

(a) The angle of projection for a particular throw is 40°.

(i) (A) State the release speed of the shot at this angle. 1

(B) Calculate the horizontal component of the initial velocity of the shot. 1

Space for working and answer

(C) Calculate the vertical component of the initial velocity of the shot. 1

Space for working and answer

(ii) The maximum height reached by the shot is 4·7 m above the ground. The time between release and reaching this height is 0·76 s.

(A) Calculate the total time between the shot being released and hitting the ground at X. 4

Space for working and answer

MARKS | DO NOT WRITE IN THIS MARGIN

1. (a) (ii) **(continued)**

(B) Calculate the range of the shot for this throw. **3**

Space for working and answer

(b) Using information from the graph, explain the effect of increasing the angle of projection on the kinetic energy of the shot at release. **2**

MARKS | DO NOT WRITE IN THIS MARGIN

2. A student sets up an experiment to investigate collisions between two trolleys on a long, horizontal track.

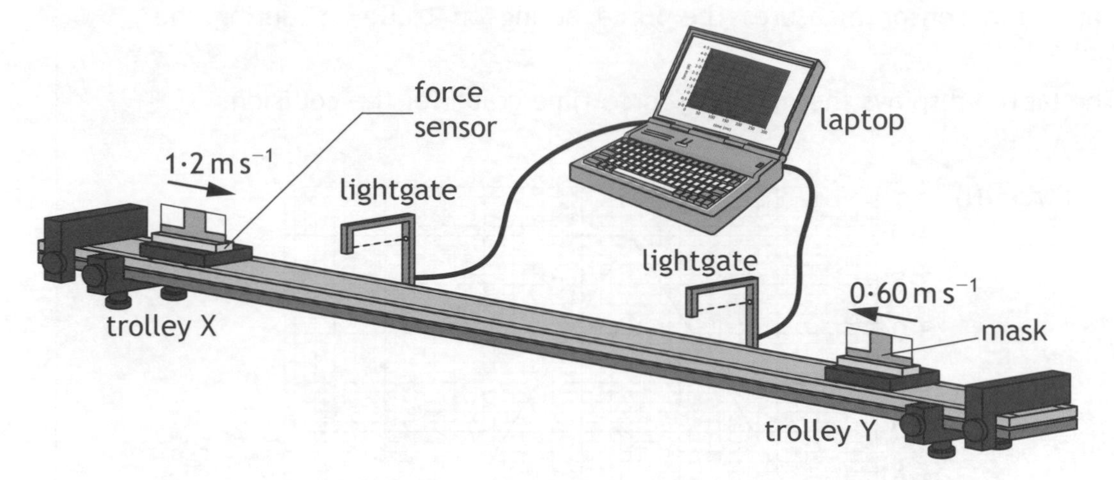

The mass of trolley X is 0·25 kg and the mass of trolley Y is 0·45 kg.

The effects of friction are negligible.

In one experiment, trolley X is moving at $1{\cdot}2\,\text{m}\,\text{s}^{-1}$ to the right and trolley Y is moving at $0{\cdot}60\,\text{m}\,\text{s}^{-1}$ to the left.

The trolleys collide and do not stick together. After the collision, trolley X rebounds with a velocity of $0{\cdot}80\,\text{m}\,\text{s}^{-1}$ to the left.

(a) Determine the velocity of trolley Y after the collision. **3**

Space for working and answer

[Turn over

MARKS | DO NOT WRITE IN THIS MARGIN

2. (continued)

(b) The force sensor measures the force acting on trolley Y during the collision.

The laptop displays the following force-time graph for the collision.

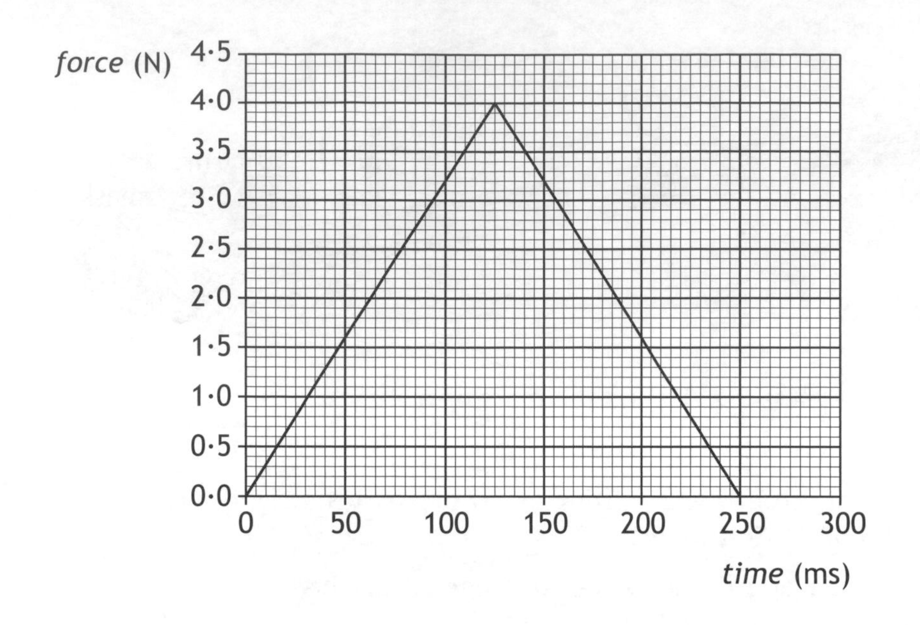

(i) Determine the magnitude of the impulse on trolley Y. **3**

Space for working and answer

(ii) Determine the magnitude of the change in momentum of trolley X. **1**

MARKS | DO NOT WRITE IN THIS MARGIN

2. (b) (continued)

(iii) Sketch a velocity-time graph to show how the velocity of trolley X varies from 0·50 s before the collision to 0·50 s after the collision. **3**

Numerical values are required on both axes.
You may wish to use the square-ruled paper on *Page thirty-six*.

[Turn over

MARKS | DO NOT WRITE IN THIS MARGIN

3. A space probe of mass $5{\cdot}60 \times 10^3$ kg is in orbit at a height of $3{\cdot}70 \times 10^6$ m above the surface of Mars.

not to scale

The mass of Mars is $6{\cdot}42 \times 10^{23}$ kg.
The radius of Mars is $3{\cdot}39 \times 10^6$ m.

(a) Calculate the gravitational force between the probe and Mars. **3**

Space for working and answer

(b) Calculate the gravitational field strength of Mars at this height. **3**

Space for working and answer

[Turn over for Question 4 on *Page fourteen*

DO NOT WRITE ON THIS PAGE

MARKS | DO NOT WRITE IN THIS MARGIN

4. Light from the Sun is used to produce a visible spectrum.

A student views this spectrum and observes a number of dark lines as shown.

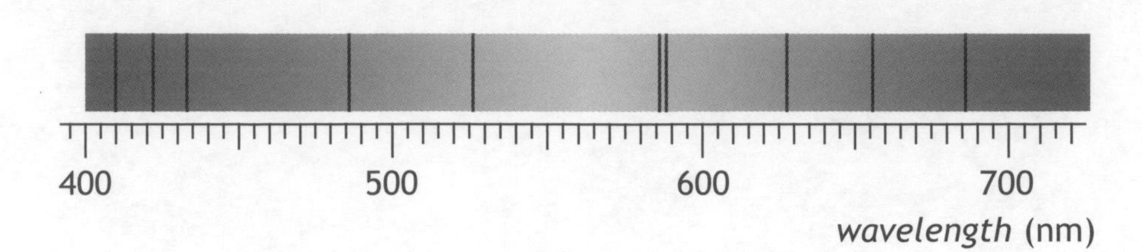

(a) Explain how these dark lines in the spectrum of sunlight are produced. **2**

(b) One of the lines is due to hydrogen.

The position of this hydrogen line in the visible spectrum is shown for a distant galaxy, a nearby galaxy and the Sun.

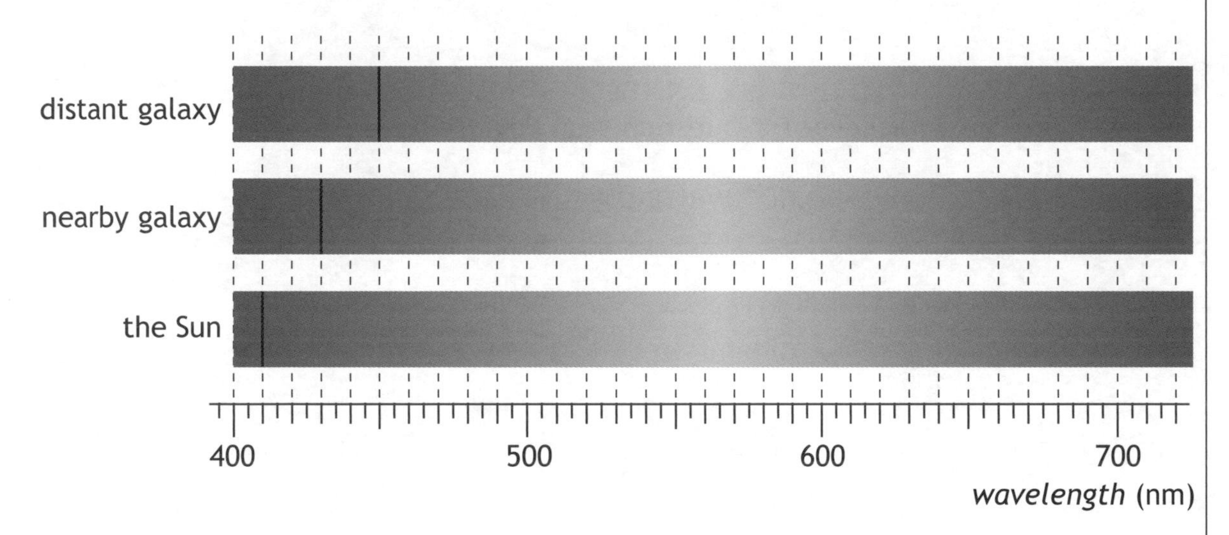

(i) Explain why the position of the line is different in each of the spectra. **2**

MARKS DO NOT WRITE IN THIS MARGIN

4. (b) (continued)

(ii) Show that the redshift of the light from the distant galaxy is 0·098. **2**

Space for working and answer

(iii) Calculate the approximate distance to the distant galaxy. **5**

Space for working and answer

[Turn over

MARKS DO NOT WRITE IN THIS MARGIN

5. A quote from a well-known science fiction writer states:

"In the beginning there was nothing, which exploded."

Using your knowledge of physics, comment on the above statement. **3**

MARKS | DO NOT WRITE IN THIS MARGIN

6. (a) The Standard Model classifies *force mediating particles* as bosons. Name the boson associated with the electromagnetic force. **1**

(b) In July 2012 scientists at CERN announced that they had found a particle that behaved in the way that they expected the Higgs boson to behave. Within a year this particle was confirmed to be a Higgs boson.

This Higgs boson had a mass-energy equivalence of 126 GeV.
($1\,eV = 1{\cdot}6 \times 10^{-19}\,J$)

(i) Show that the mass of the Higgs boson is $2{\cdot}2 \times 10^{-25}\,kg$. **3**

Space for working and answer

(ii) Compare the mass of the Higgs boson with the mass of a proton in terms of orders of magnitude. **2**

Space for working and answer

[Turn over

MARKS DO NOT WRITE IN THIS MARGIN

7. The use of analogies from everyday life can help better understanding of physics concepts. Throwing different balls at a coconut shy to dislodge a coconut is an analogy which can help understanding of the photoelectric effect.

Use your knowledge of physics to comment on this analogy. **3**

[Turn over for Question 8 on *Page twenty*

DO NOT WRITE ON THIS PAGE

MARKS DO NOT WRITE IN THIS MARGIN

8. A student investigates how irradiance I varies with distance d from a point source of light.

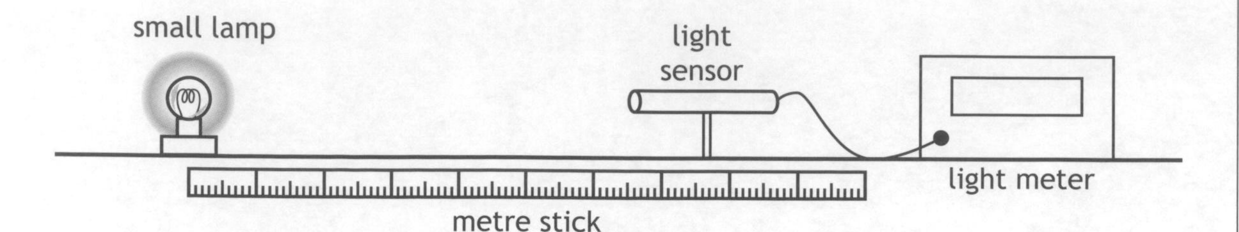

The distance between a small lamp and a light sensor is measured with a metre stick. The irradiance is measured with a light meter.

The apparatus is set up as shown in a darkened laboratory.

The following results are obtained.

d (m)	0·20	0·30	0·40	0·50
I ($W\,m^{-2}$)	134·0	60·5	33·6	21·8

(a) State what is meant by the term *irradiance*. **1**

(b) Use **all** the data to establish the relationship between irradiance I and distance d. **3**

MARKS | DO NOT WRITE IN THIS MARGIN

8. (continued)

(c) The lamp is now moved to a distance of 0·60 m from the light sensor.

Calculate the irradiance of light from the lamp at this distance. **3**

Space for working and answer

(d) Suggest one way in which the experiment could be improved.

You **must** justify your answer. **2**

(e) The student now replaces the lamp with a different small lamp.
The power output of this lamp is 24 W.

Calculate the irradiance of light from this lamp at a distance of 2·0 m. **4**

Space for working and answer

MARKS | DO NOT WRITE IN THIS MARGIN

9. A student carries out two experiments to investigate the spectra produced from a ray of white light.

(a) In the first experiment, a ray of white light is incident on a glass prism as shown.

not to scale

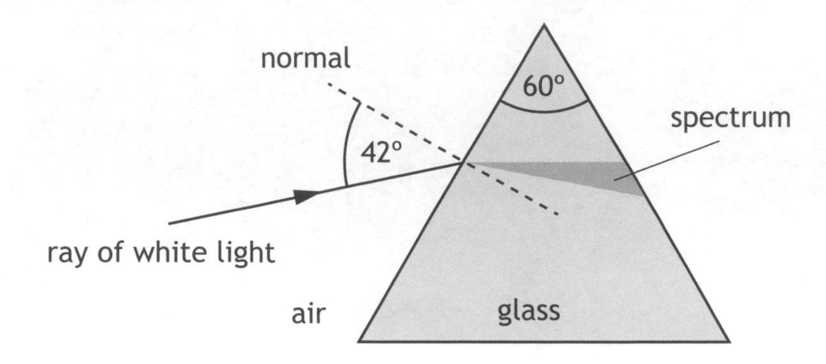

(i) Explain why a spectrum is produced in the glass prism. **1**

(ii) The refractive index of the glass for red light is 1·54.

Calculate the speed of red light in the glass prism. **3**

Space for working and answer

MARKS DO NOT WRITE IN THIS MARGIN

9. (continued)

(b) In the second experiment, a ray of white light is incident on a grating.

not to scale

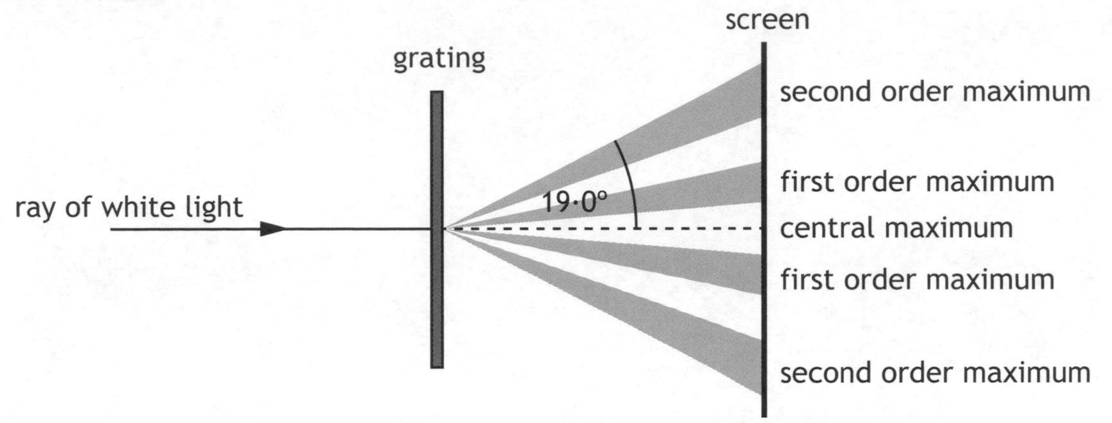

The angle between the central maximum and the second order maximum for red light is 19·0°.
The frequency of this red light is $4{\cdot}57 \times 10^{14}$ Hz.

(i) Calculate the distance between the slits on this grating. **5**

Space for working and answer

(ii) Explain why the angle to the second order maximum for blue light is different to that for red light. **3**

[BLANK PAGE]

DO NOT WRITE ON THIS PAGE

MARKS DO NOT WRITE IN THIS MARGIN

10. A car battery is connected to an electric motor as shown.

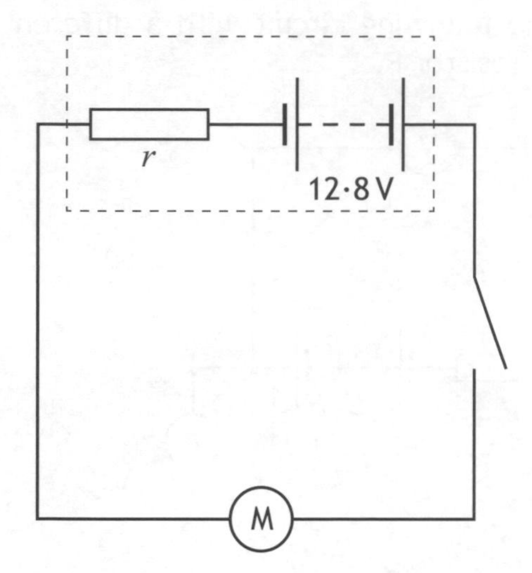

The electric motor requires a large current to operate.

(a) The car battery has an e.m.f. of 12·8 V and an internal resistance r of $6{\cdot}0 \times 10^{-3}\,\Omega$. The motor has a resistance of $0{\cdot}050\,\Omega$.

(i) State what is meant by an *e.m.f. of 12·8 V*. **1**

(ii) Calculate the current in the circuit when the motor is operating. **3**

Space for working and answer

(iii) Suggest why the connecting wires used in this circuit have a large diameter. **1**

MARKS DO NOT WRITE IN THIS MARGIN

10. (continued)

(b) A technician sets up the following circuit with a different car battery connected to a variable resistor R.

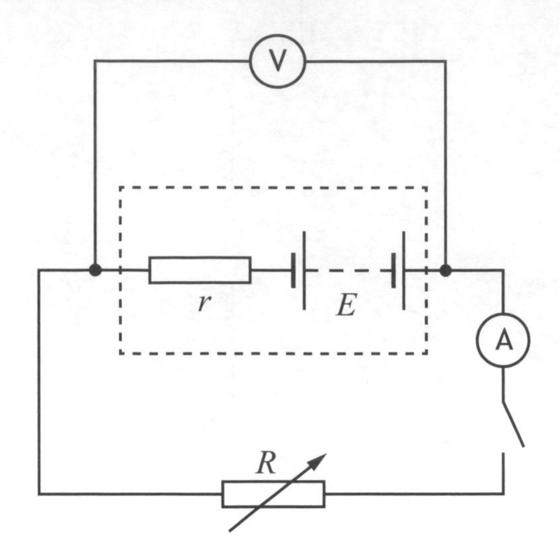

Readings of current I and terminal potential difference V from this circuit are used to produce the following graph.

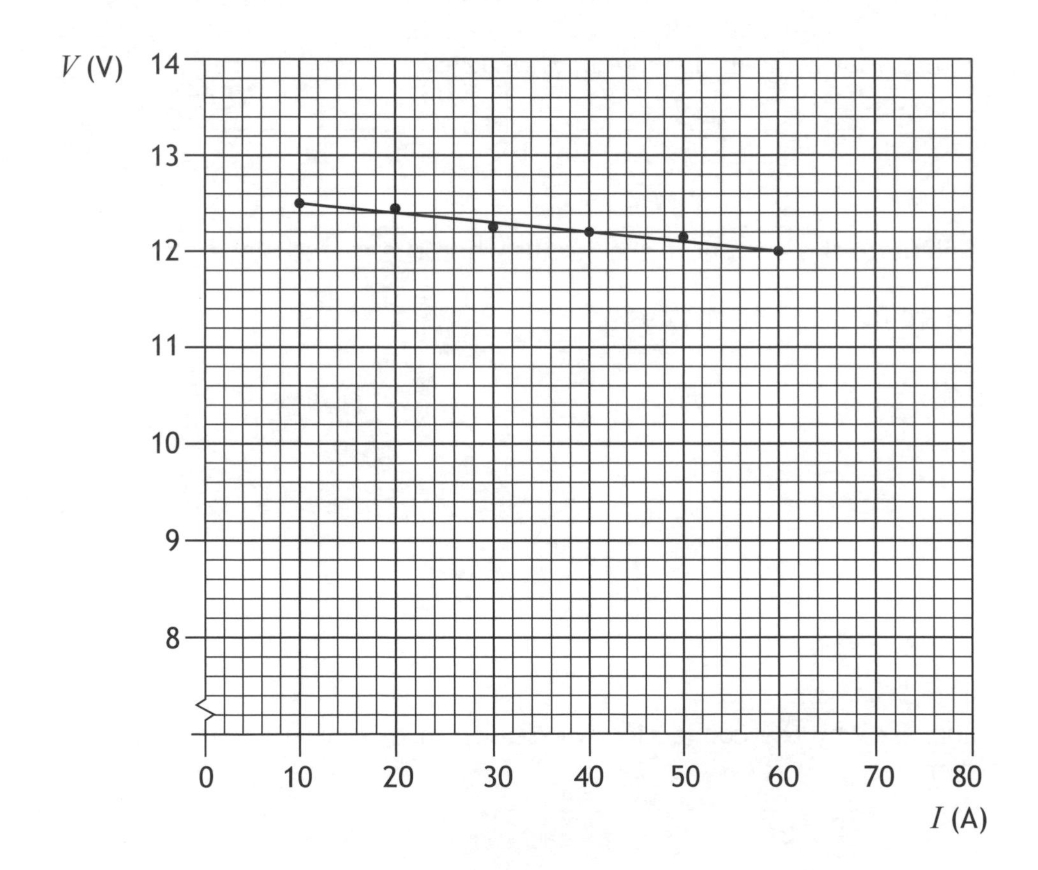

MARKS DO NOT WRITE IN THIS MARGIN

10. (b) (continued)

Use information from the graph to determine:

(i) the e.m.f. of the battery; 1

Space for working and answer

(ii) the internal resistance of the battery; 3

Space for working and answer

[Turn over

MARKS | DO NOT WRITE IN THIS MARGIN

10. (b) (continued)

(iii) After being used for some time the e.m.f. of the battery decreases to 11·5 V and the internal resistance increases to 0·090 Ω.

The battery is connected to a battery charger of constant e.m.f. 15·0 V and internal resistance of 0·45 Ω as shown.

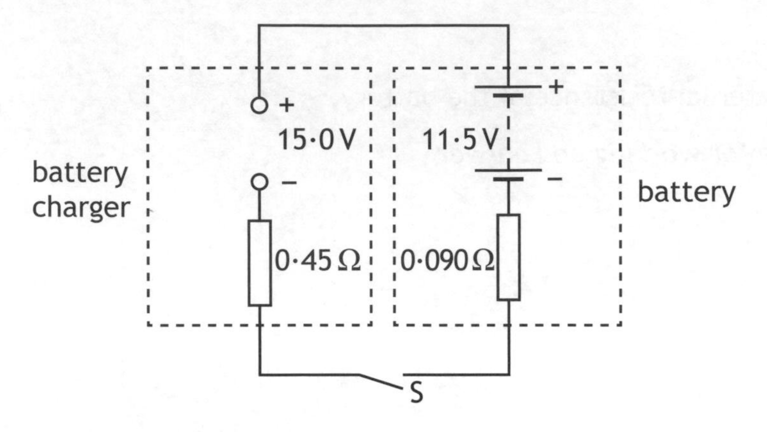

(A) Switch S is closed.

Calculate the initial charging current. **3**

Space for working and answer

(B) Explain why the charging current decreases as the battery charges. **2**

[Turn over for Question 11 on *Page thirty*]

DO NOT WRITE ON THIS PAGE

MARKS DO NOT WRITE IN THIS MARGIN

11. A defibrillator is a device that provides a high energy electrical impulse to correct abnormal heart beats.

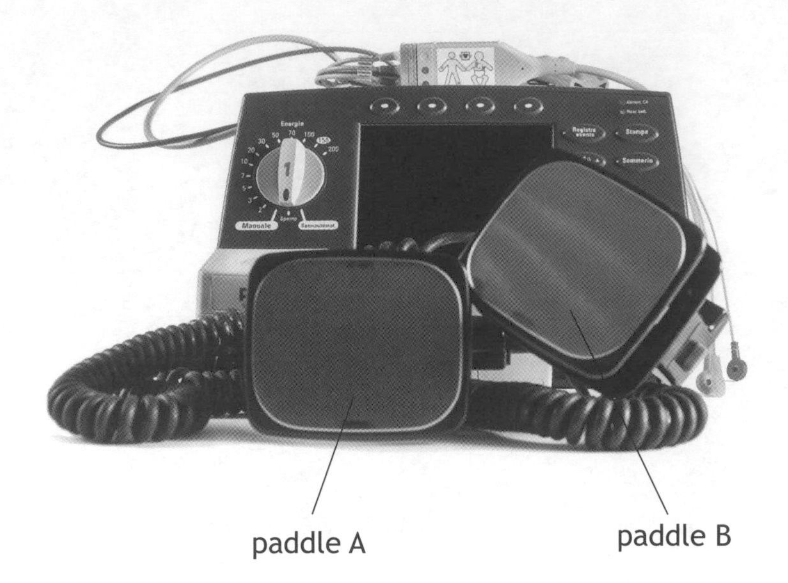

The diagram shows a simplified version of a defibrillator circuit.

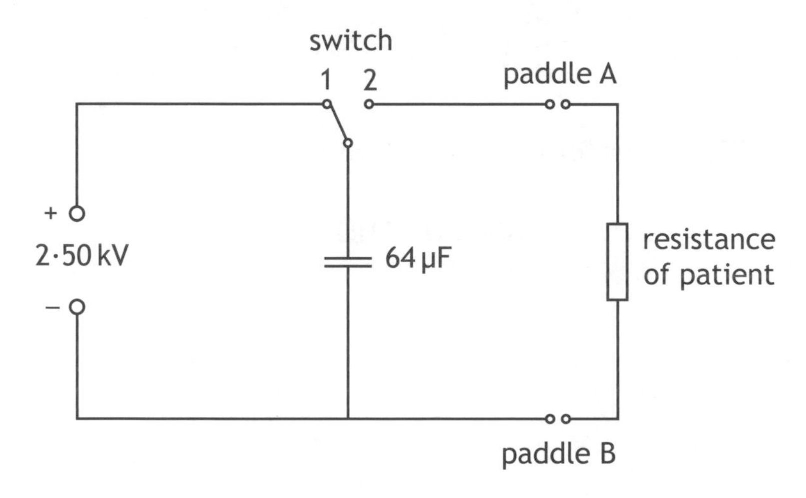

The switch is set to position 1 and the capacitor charges.

(a) Show the charge on the capacitor when it is fully charged is 0·16 C. **2**

Space for working and answer

MARKS | DO NOT WRITE IN THIS MARGIN

11. (continued)

(b) Calculate the maximum energy stored by the capacitor. **3**

Space for working and answer

(c) To provide the electrical impulse required the capacitor is discharged through the person's chest using the paddles as shown

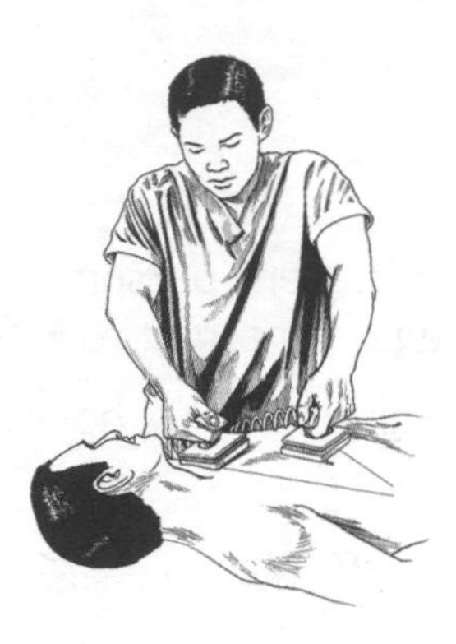

The initial discharge current through the person is 35·0 A.

(i) Calculate the effective resistance of the part of the person's body between the paddles. **3**

Space for working and answer

MARKS | DO NOT WRITE IN THIS MARGIN

11. (c) (continued)

(ii) The graph shows how the current between the paddles varies with time during the discharge of the capacitor.

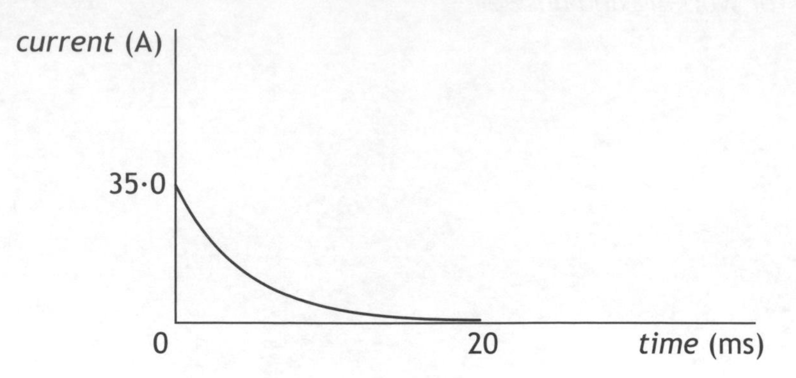

The effective resistance of the person remains the same during this time.

Explain why the current decreases with time. **1**

(iii) The defibrillator is used on a different person with larger effective resistance. The capacitor is again charged to 2·50 kV.

On the graph in (c)(ii) add a line to show how the current in this person varies with time.

(An additional graph, if required, can be found on *Page thirty-eight*). **2**

MARKS DO NOT WRITE IN THIS MARGIN

12. A student carries out an investigation to determine the refractive index of a prism.

A ray of monochromatic light passes through the prism as shown.

not to scale

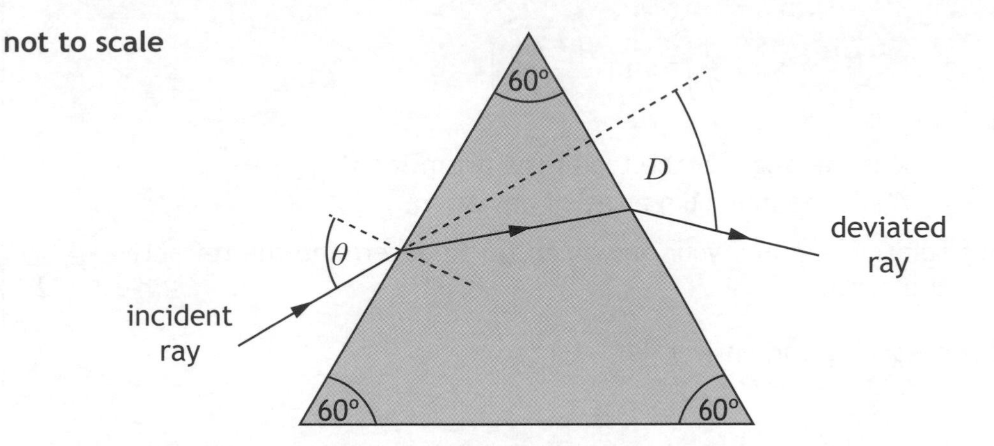

The angle of deviation D is the angle between the direction of the incident ray and the deviated ray.

The student varies the angle of incidence θ and measures the corresponding angles of deviation D.

The results are shown in the table.

Angle of incidence θ (°)	*Angle of deviation* D (°)
30·0	47·0
40·0	38·1
50·0	37·5
60·0	38·8
70·0	42·5

(a) Using the square-ruled paper on *Page thirty-five*, draw a graph of D against θ. **3**

(b) Using your graph state the two values of θ that produce an angle of deviation of 41·0°. **1**

(c) Using your graph give an estimate of the minimum angle of deviation D_m. **1**

MARKS DO NOT WRITE IN THIS MARGIN

12. (continued)

(d) The refractive index n of the prism can be determined using the relationship.

$$n\sin\left(\frac{A}{2}\right)=\sin\left(\frac{A+D_m}{2}\right)$$

where A is the angle at the top of the prism, and
D_{m} is the minimum angle of deviation.

Use this relationship and your answer to (c) to determine the refractive index of the prism. **2**

Space for working and answer

(e) Using the same apparatus, the student now wishes to determine more precisely the minimum angle of deviation.

Suggest two improvements to the experimental procedure that would achieve this. **2**

[END OF QUESTION PAPER]

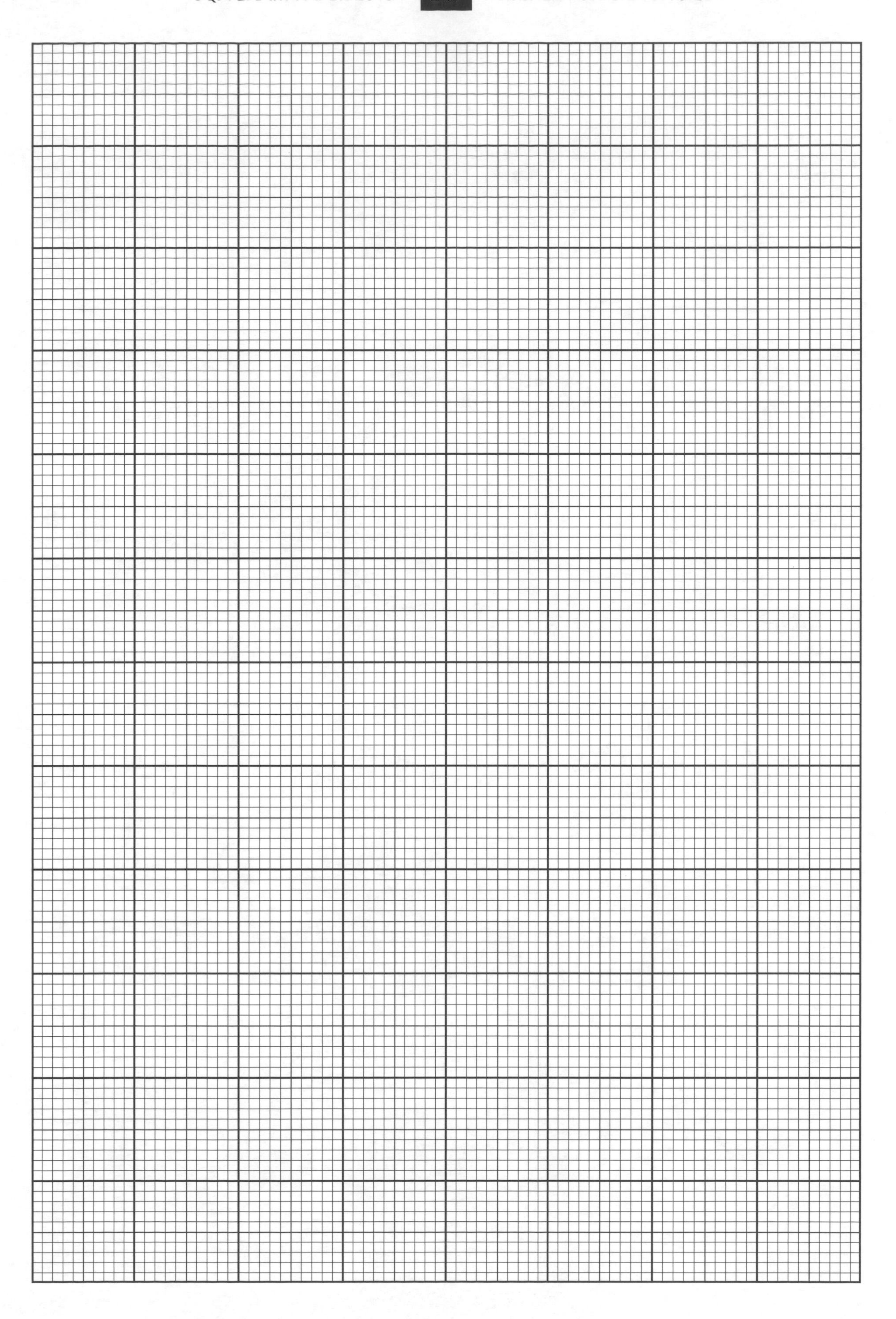

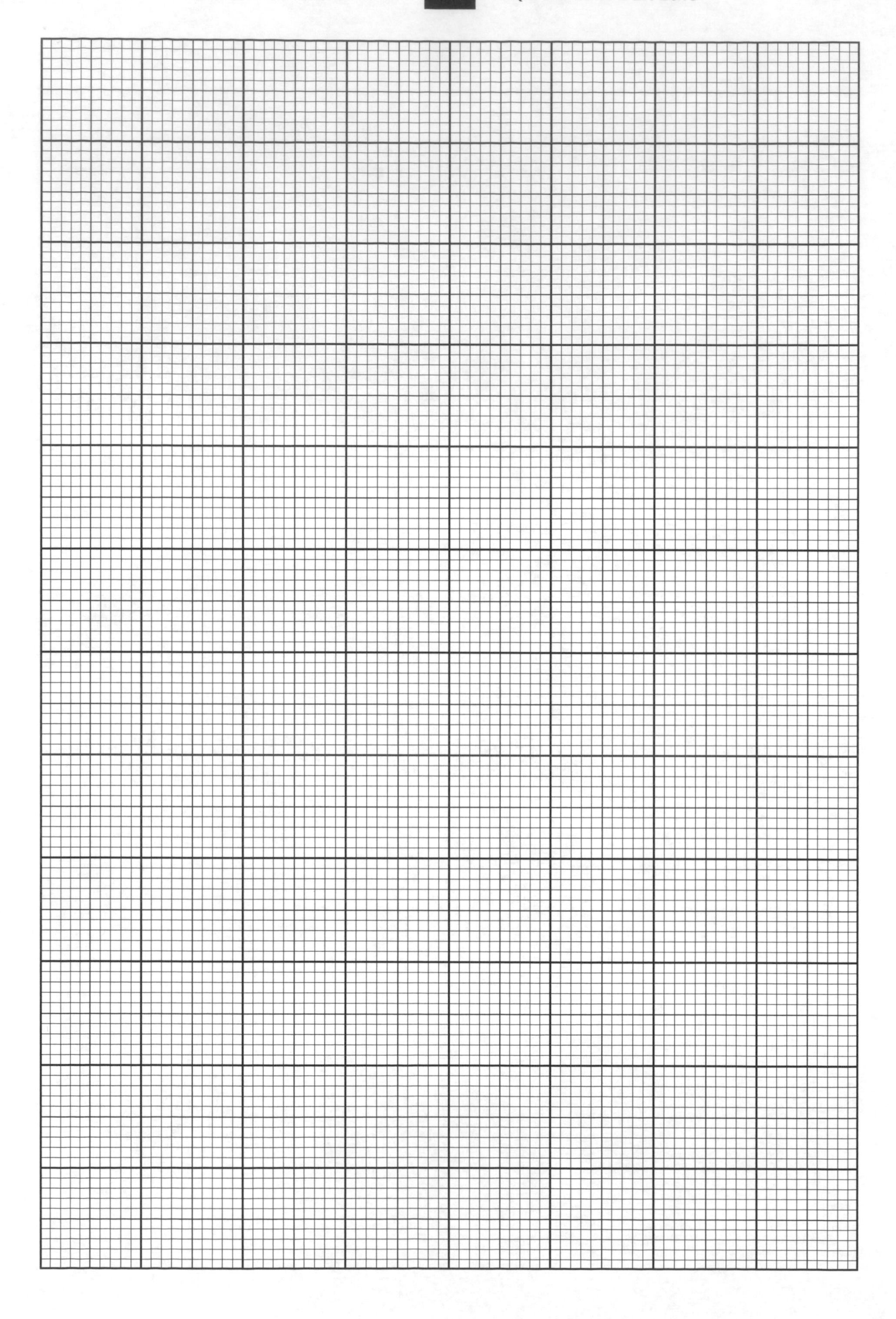

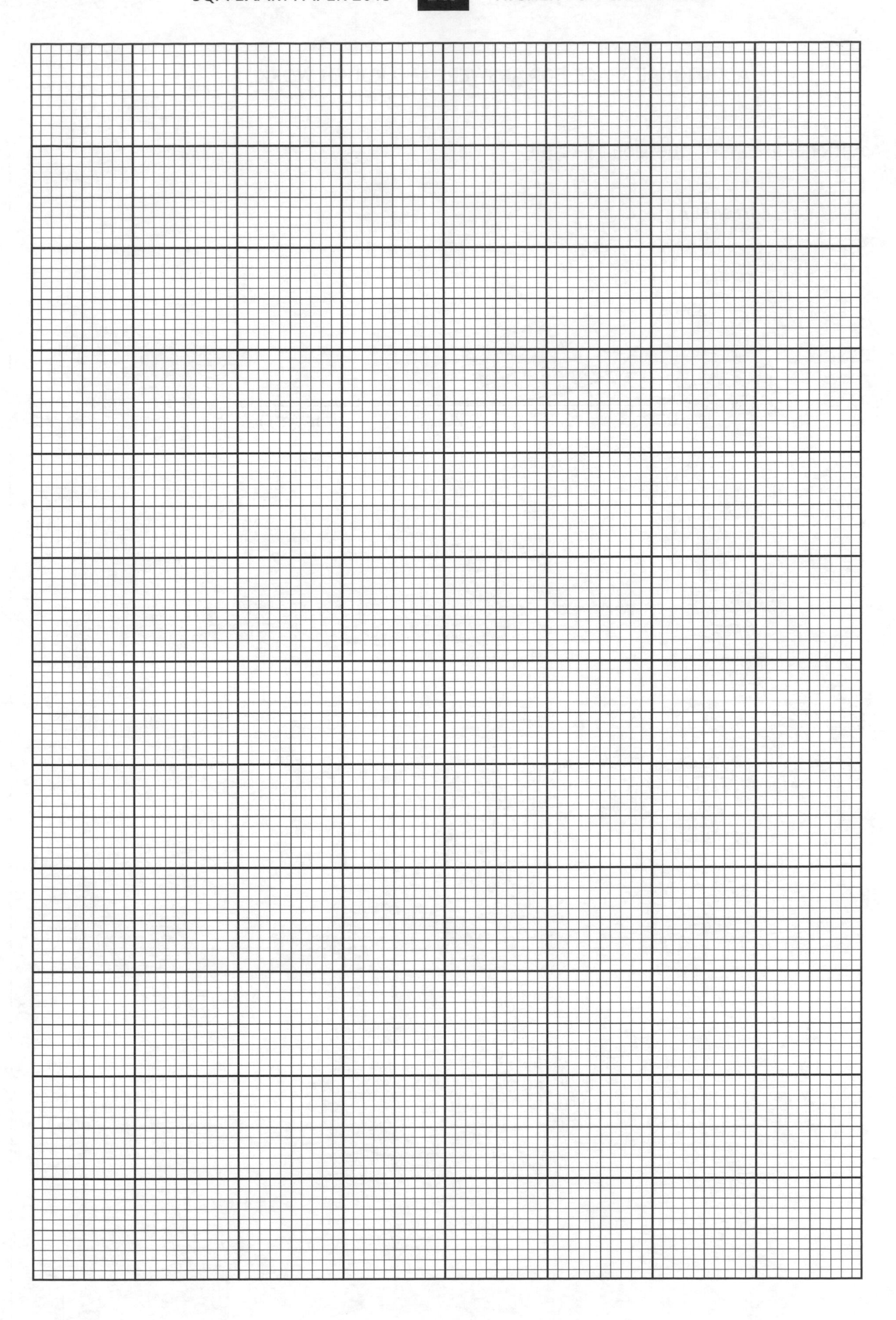

MARKS | DO NOT WRITE IN THIS MARGIN

ADDITIONAL SPACE FOR ANSWERS AND ROUGH WORK

Additional graph for Question 11 (c)(iii)

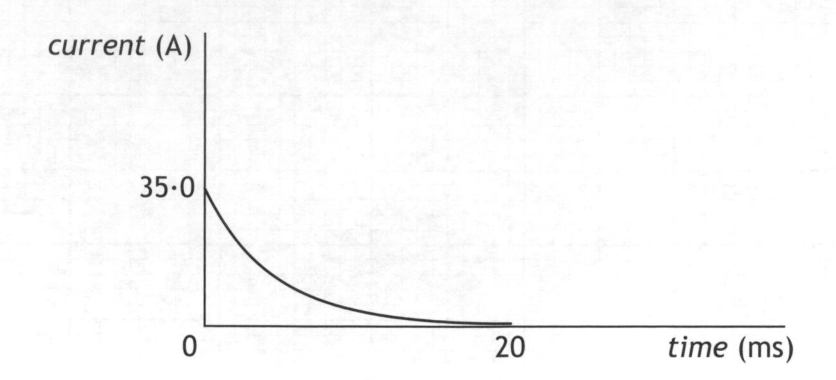

MARKS DO NOT WRITE IN THIS MARGIN

ADDITIONAL SPACE FOR ANSWERS AND ROUGH WORK

MARKS DO NOT WRITE IN THIS MARGIN

ADDITIONAL SPACE FOR ANSWERS AND ROUGH WORK

HIGHER FOR CfE | ANSWER SECTION

SQA AND HODDER GIBSON HIGHER FOR CfE PHYSICS 2015

HIGHER FOR CfE PHYSICS SPECIMEN QUESTION PAPER

SECTION 1

Question	Answer	Mark
1.	E	1
2.	A	1
3.	D	1
4.	B	1
5.	A	1
6.	B	1
7.	E	1
8.	C	1
9.	C	1
10.	A	1
11.	E	1
12.	A	1
13.	C	1
14.	A	1
15.	E	1
16.	D	1
17.	E	1
18.	C	1
19.	A	1
20.	B	1

SECTION 2

Question			Answer	Max mark
1.	(a)	(i)	(Initial horizontal component = $v\cos\theta$ = 50 cos35) = 41 ms^{-1} (1)	1
		(ii)	(Initial vertical component = $v\sin\theta$ = 50 sin 35) = 29 ms^{-1} (1)	1
	(b)		$v = u + at$ (1) v = 29 − 9·8t (1) t = (0 − 29)/−9·8 = 2·96 (s) t_{TOTAL} = 5·92 (s) (1) $d_h = v_h t$ = 41 × 5·92 (1) (= 240 m)	4
2.	(a)	(i)	Component of weight down slope = $mg\sin\theta$ (1) = 220 × 9·8 × *sin* 3·2° (1) (=120 N)	2
		(ii)	Unbalanced Force = 230 − (120+48) = 62 N (1) $F = ma$ (1) 62 = 220 × a (1) a = 0·28 m s^{-2} (1)	4
		(iii)	As angle (of slope) increases $mg\sin\theta$ increases (1) When $mg\sin\theta \geqq$ engine force − friction, the vehicle cannot move up the slope (1)	2
	(b)	(i)	lost volts = Ir (1) = 22 × 0·52 (1) (= 11 V)	2
		(ii)	*p.d.* = 48 − 11 = 37 V (1) P = I V (1) = 22 × 37 (1) = 810 W (1)	4
	(c)		terminal potential difference decreases (1) current increases (1) lost volts increases (1)	3
3.			estimate of masses (500 kg < car mass < 3000 kg) (1) estimate of speed (10 ms^{-1} < speed < 70 ms^{-1}) (1) $E_k = 1/2\ mv^2$ (1) Final answer and unit (1)	4
4.	(a)		$d = vt$ (1) $d = (3 \times 10^8 \times 0{\cdot}995) \times 2{\cdot}2 \times 10^6$ (1) d = 660 m	2
	(b)		$t' = \dfrac{t}{\sqrt{1-\left(\frac{v}{c}\right)^2}}$ (1) $t' = \dfrac{2{\cdot}2 \times 10^{-6}}{\sqrt{1-\left(\frac{0{\cdot}995}{1}\right)^2}}$ (1) $= 2{\cdot}2 \times 10^{-5}$ s (1)	3
	(c)		For an observer on Earth's frame of reference the mean life of the muon is much greater **OR** The distance in the muon frame of reference is shorter	1

Question			Answer	Max mark
5.			This open-ended question requires comment on the **suitability of the design of the bicycle helmet**. Candidate responses are expected to make judgements on its suitability, on the basis of relevant physics ideas/concepts which might include one or more of: 'crumple zone'; impulse; energy being absorbed; air circulation and aerodynamics; or other relevant ideas/concepts.	
			3 marks: The candidate has demonstrated a **good** conceptual understanding of the physics involved, providing a logically correct response to the problem/situation presented. This type of response might include a statement of principle(s) involved, a relationship or equation, and the application of these to respond to the problem/situation. This does not mean the answer has to be what might be termed an 'excellent' answer or a 'complete' one. In response to this question, a **good** understanding might be demonstrated by a candidate response that: • makes a judgement on suitability based on one relevant physics idea/concept, in a **detailed/developed** response that is **correct or largely correct** (any weaknesses are minor and do not detract from the overall response), **OR** • makes judgement(s) on suitability based on a range of relevant physics ideas/concepts, in a response that is **correct or largely correct** (any weaknesses are minor and do not detract from the overall response), **OR** • otherwise demonstrates a good understanding of the physics involved.	
			2 marks: The candidate has demonstrated a **reasonable** understanding of the physics involved, showing that the problem/situation is understood. This type of response might make some statement(s) that is/are relevant to the problem/situation, for example, a statement of relevant principle(s) or identification of a relevant relationship or equation. In response to this question, a **reasonable** understanding might be demonstrated by a candidate response that: • makes a judgement on suitability based on one or more relevant physics idea(s)/concept(s), in a response that is **largely correct** but has **weaknesses** which detract to a small extent from the overall response, **OR** • otherwise demonstrates a reasonable understanding of the physics involved.	
			1 mark: The candidate has demonstrated a **limited** understanding of the physics involved, showing that a little of the physics that is relevant to the problem/situation is understood. The candidate has made some statement(s) that is/are relevant to the problem/situation. In response to this question, a **limited** understanding might be demonstrated by a candidate response that: • makes a judgement on suitability based on one or more relevant physics idea(s)/concept(s), in a response that has **weaknesses** which detract to a large extent from the overall response, **OR** • otherwise demonstrates a limited understanding of the physics involved.	

Question			Answer	Max mark
			0 marks: The candidate has demonstrated **no** understanding of the physics that is relevant to the problem/situation. The candidate has made no statement(s) that is/are relevant to the problem/situation. Where the candidate has only demonstrated knowledge and understanding of physics **that is not relevant to the problem/situation presented**, 0 marks should be awarded.	
6.	(a)	(i)	The star is moving away from the Earth (1) Plus any one point from the following for 1 mark: • The apparent wavelength of the hydrogen spectra from the star has increased • The apparent frequency of the hydrogen spectra from the star is less than the actual frequency on Earth • The frequency of the light from the star has shifted towards the red end of the spectrum • Light from the star is experiencing a Doppler shift.	2
		(ii)	$z = \frac{(\lambda_{obs} - \lambda_{rest})}{\lambda_{rest}}$ (1) $z = \frac{(676 \times 10^{-9} - 656 \times 10^{-9})}{656 \times 10^{-9}}$ (1) $z = 0{\cdot}03$ (1) $z = \frac{v}{c}$ (1) $v = 0{\cdot}03c$ $\underline{v = 9 \times 10^{6}\ \text{m s}^{-1}}$ (1)	5
	(b)		$v = H_0 d$ (1) $d = \frac{v}{H_0}$ $d = \frac{1{\cdot}2 \times 10^{7}}{2{\cdot}3 \times 10^{-18}}$ (1) $d = 5{\cdot}2 \times 10^{24}$ m	2

Question			Answer	Max mark
	(c)		This open-ended question requires comment on the **suitability of the expanding balloon model to explain the expansion of the universe**. Candidate responses are expected to make judgements on its suitability, on the basis of relevant physics ideas/concepts which might include one or more of: that distances between the dots increase similarly as the distances between the galaxies; it is the 2-dimensional surface that is being compared to 3-dimensional space – so centre of balloon has no physical analogue; galaxies themselves do not expand – they are bound by gravitation; or other relevant ideas/concepts.	
			3 marks: The candidate has demonstrated a **good** conceptual understanding of the physics involved, providing a logically correct response to the problem/situation presented. This type of response might include a statement of principle(s) involved, a relationship or equation, and the application of these to respond to the problem/situation. This does not mean the answer has to be what might be termed an 'excellent' answer or a 'complete' one. In response to this question, a **good** understanding might be demonstrated by a candidate response that: • makes a judgement on suitability based on one relevant physics idea/concept, in a **detailed/developed** response that is **correct or largely correct** (any weaknesses are minor and do not detract from the overall response), **OR** • makes judgement(s) on suitability based on a range of relevant physics ideas/concepts, in a response that is **correct or largely correct** (any weaknesses are minor and do not detract from the overall response), **OR** • otherwise demonstrates a good understanding of the physics involved.	

Question			Answer	Max mark
			2 marks: The candidate has demonstrated a **reasonable** understanding of the physics involved, showing that the problem/situation is understood. This type of response might make some statement(s) that is/are relevant to the problem/situation, for example, a statement of relevant principle(s) or identification of a relevant relationship or equation. In response to this question, a **reasonable** understanding might be demonstrated by a candidate response that: • makes a judgement on suitability based on one or more relevant physics idea(s)/concept(s), in a response that is **largely correct** but has **weaknesses** which detract to a small extent from the overall response, **OR** • otherwise demonstrates a reasonable understanding of the physics involved.	
			1 mark: The candidate has demonstrated a **limited** understanding of the physics involved, showing that a little of the physics that is relevant to the problem/situation is understood. The candidate has made some statement(s) that is/are relevant to the problem/situation. In response to this question, a **limited** understanding might be demonstrated by a candidate response that: • makes a judgement on suitability based on one or more relevant physics idea(s)/concept(s), in a response that has **weaknesses** which detract to a large extent from the overall response, **OR** • otherwise demonstrates a limited understanding of the physics involved.	
			0 marks: The candidate has demonstrated **no** understanding of the physics that is relevant to the problem/situation. The candidate has made no statement(s) that is/are relevant to the problem/situation. Where the candidate has only demonstrated knowledge and understanding of physics **that is not relevant to the problem/situation presented**, 0 marks should be awarded.	
7.	(a)	(i)	A = 2u + 1d	1
			B = 1u + 2d	1
		(ii)	gluon	1
	(b)		beta decay	1
8.	(a)	(i)	$W = QV$ or $E_w = QV$ (1) $E_w = 1{\cdot}6 \times 10^{-19} \times 35000$ (1) $E_w = 5{\cdot}6 \times 10^{-15}$ J	2
		(ii)	Original $Ek = \frac{1}{2}mv^2$ (1) $E_k = \frac{1}{2}(1{\cdot}673 \times 10^{-27})(1{\cdot}2 \times 10^6)^2$ (1) $E_k = 1{\cdot}20 \times 10^{-15}$ (J) New $E_k = 1{\cdot}20 \times 10^{-15} + 5{\cdot}6 \times 10^{-15}$ (J) New $E_k = 6{\cdot}8 \times 10^{-15}$ (J) (1) $E_k = \frac{1}{2}mv^2$ $6{\cdot}8 \times 10^{-15} = \frac{1}{2}(1{\cdot}673 \times 10^{-27})v^2$ (1) $v = 2{\cdot}9 \times 10^6$ m s^{-1} (1)	5
	(b)		Alternating voltage has constant frequency (1) **OR** As speed of protons increases, they travel further in the same time. (1)	1
9.	(a)	(i)	$\Delta m = 4 \times 1{\cdot}673 \times 10^{-27} - 6{\cdot}646 \times 10^{-27}$ $\Delta m = 4{\cdot}6 \times 10^{-29}$ (kg) (1) $E = mc^2$ (1) $E = 4{\cdot}6 \times 10^{-29} \times (3{\cdot}00 \times 10^8)^2$ (1) $E = 4{\cdot}14 \times 10^{-12}$ J (1)	4
		(ii)	1 kg hydrogen has $\frac{0{\cdot}20}{1{\cdot}673 \times 10^{-27}} = 1{\cdot}195 \times 10^{26}$ atoms (1) Provides $\frac{1{\cdot}195 \times 10^{26}}{4} = 0{\cdot}2989 \times 10^{26}$ reactions (1) Releases $0{\cdot}2989 \times 10^{26} \times 4{\cdot}14 \times 10^{-12}$ $= 1{\cdot}2 \times 10^{14}$ J (1)	3
		(iii)	Large amount of energy released results in very high temperatures **OR** Strong magnetic fields are required for containment	1

Question			Answer	Max mark
	(b)		$m_{Rn}v_{Rn} = -m_a v_a$ (1) $3{\cdot}653 \times 10^{-25} \times v_{Rn} = 6{\cdot}645 \times 10^{-27} \times 1{\cdot}46 \times 10^7$ (1) $v_{Rn} = 2{\cdot}656 \times 10^5$ m s^{-1} (1)	3
10.	(a)		Blue light has higher frequency/ energy per photon than red light. (1) Photons of red light do not have enough energy to eject electrons (1)	2
	(b)		$E_k = hf - hf_0$ (1) $= (6{\cdot}63 \times 10^{-34} \times 7{\cdot}0 \times 10^{14}) - 2{\cdot}0 \times 10^{-19}$ (1) $= 2{\cdot}6 \times 10^{-19}$ J (1)	3
11.	(a)		Light with fixed/no phase difference.	1
	(b)	(i)	Bright fringes are produced by waves meeting in phase/crest to crest/trough to trough	1
		(ii)	$\Delta x = \frac{\lambda D}{d}$ (1) $\frac{9{\cdot}5 \times 10^{-3}}{4} = \frac{633 \times 10^{-9} \times 0{\cdot}750}{d}$ (1) $d = 2{\cdot}0 \times 10^{-4}$ m (1)	4
		(iii)	$\%uncert\Delta x = \frac{0{\cdot}2 \times 100}{9{\cdot}5 \times 10^{-3}} = 2{\cdot}1\%$ (1) $\frac{0{\cdot}002 \times 100}{0{\cdot}750} = 0{\cdot}27\%$ (1) Improve precision in measurement of Δx (1)	3
		(iv)	Green laser ➔ shorter λ (1) Fringes closer together (1)	2
12.	(a)	(i)	Labels (quantities and units) and scale (1) Points correctly plotted (1) Correct best fit line (1)	3
		(ii)	Gradient of graph (1) Refractive index = 1·50 (1)	2
		(iii)	Repeated measurements Increased range of measurements Narrower beam of light Increase the number of values within the range Protractor with more precise scale eg ½° divisions	2
	(b)		$\sin\theta_c = \frac{1}{n}$ (1) $\theta_c = \sin^{-1}\frac{1}{1{\cdot}54}$ (1) $\theta_c = 40{\cdot}5°$ (1)	3
13.	(a)		$R = V/I$ (1) $= 12 / (30 \times 10^{-6})$ (1) $= 400\ 000\ \Omega$ (1)	3
	(b	(i)	$Q = It$ (1) $= 30 \times 10^{-6} \times 30$ (1) $= 900 \times 10^{-6}$ C (1)	3
		(ii)	$C = Q/V$ (1) $200 \times 10^{-6} = 900 \times 10^{-6} / V$ (1) $V = 4{\cdot}5$ V (1) Therefore voltage across resistor is $12 - 4{\cdot}5 = 7{\cdot}5$ V (1)	4
14.	(a)		Material 2	1
	(b)		resistance decreases (1) electron jumps (from valence band) to conduction band (1)	2

HIGHER FOR CfE PHYSICS MODEL PAPER 1

SECTION 1

Question	Answer	Mark
1.	A	1
2.	E	1
3.	B	1
4.	C	1
5.	B	1
6.	B	1
7.	D	1
8.	D	1
9.	D	1
10.	E	1
11.	D	1
12.	A	1
13.	E	1
14.	B	1
15.	C	1
16.	B	1
17.	B	1
18.	C	1
19.	C	1
20.	D	1

SECTION 2

Question			Answer	Max mark
1.	(a)		Use $s = ut + ½ at^2$ sub $s = h$, $u = 0$	1
	(b)		Values of t^2 – 0.016, 0.036, 0.051, 0.073, 0.090 (s^2) (2) deduct 1 mark for each incorrect value until 0.	2
	(c)		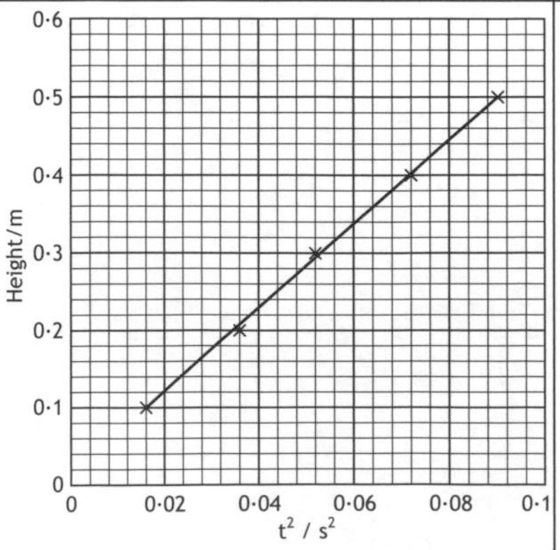 Graph: Height/m (0–0·6) against t^2 / s^2 (0–0·1) Correct points (2) Line of best fit (not forced through origin) (1) Deduct 1 max for incorrect labelling of axes (quantities or units)	3
	(d)		Gradient m = 0·5 a (1) $= \frac{y_2 - y_1}{x_2 - y1}$ (1) = 5·4 (1) a = (10·2 – 11·0) ms^{-2} (1)	4
	(e)		Increase height – use greater range (1) Increase number of repeated readings· (1)	2
2·	(a)		$t' = \frac{t}{\sqrt{(1 - \frac{v^2}{c^2})}}$ (1) $4{\cdot}8 \times 10^{-6} = \frac{2{\cdot}4 \times 10^{-6}}{\sqrt{(1 - \frac{v^2}{(3 \times 10^8)^2})}}$ (1) $v = 2{\cdot}6 \times 10^8$ ms^{-1} (1)	3
	(b)		$s = vt = 2{\cdot}6 \times 10^8 \times 4{\cdot}8 \times 10^{-6}$ = (1) $1{\cdot}25 \times 10^3$ m (2)	3
3·	(a)	(i)	$u_h = v\cos\theta$ (1) = 6·5 cos50° = 4·2 ms^{-1} (1)	2
		(ii)	$u_v = v\sin\theta$ (1) = 6·5 sin50° = 5·0 ms^{-1} (1)	2
	(b)		t = s / v (1) = 2·9 / 4·2 (1) = 0·69 s Answer given	2
	(c)		$s = ut + ½ a t^2$ (1) $= 5 \times 0{\cdot}69 + ½ \times -9{\cdot}8 \times (0{\cdot}69)^2$ = 1·1 m (1) h = 2·3 + 1·1 = 3·4 m (1)	3
	(d)		Ball would not land in basket, (1) initial vertical speed increases, so ball is higher than the basket when it has covered 2·9m horizontally (1)	2
4.	(a)	(i)	Impulse = area under F-t graph (1) = ½ x 0·010 × 70 (1) = 0·35 Ns Answer given	2
		(ii)	Change in momentum = 0·35 $kgms^{-1}$(1) upwards (1)	2
		(iii)	Impulse = mv – mu (1) 0·35 = 0·05(v – (–5·6)) (1) v = 1·4 ms^{-1} (1) **OR** ½ Ft = mv – mu (1) ½ × 70 × 0·002 = 0·05v – 0 (1) v = 1·4 ms^{-1} (1)	3
	(b)		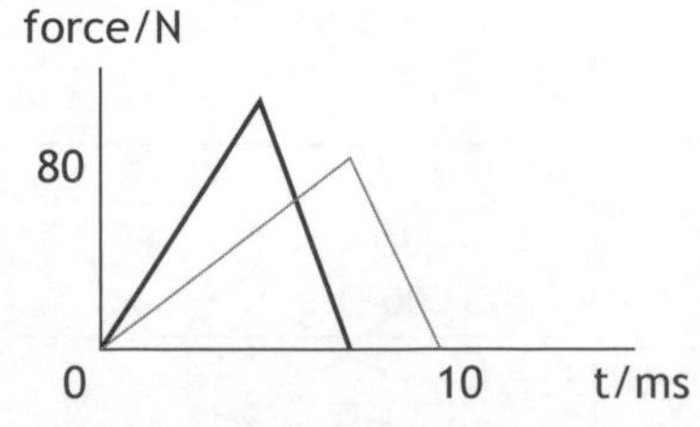 Graph: force/N (80) against t/ms (0, 10) Similar shape but, max force greater than 70N (1) max time less than 10 ms (1)	2

Question			Answer	Max mark
5.	(a)	(i)	Frequency will be greater than 294 Hz, (1) the passenger will experience more than 294 wavefronts in 1s (1)	2
		(ii)	$f_o = f_s\left(\frac{v}{v + v_s}\right)$ (1) $= f_s\left(\frac{340}{340 + 28{\cdot}0}\right)$ (1) = 272 Hz (1)	3
	(b)		Moving away (1) since wavelength increased (or frequency decreased) (1)	2
6.			**Open ended — see general marking instructions for breakdown.** Possible points: • Possibly hit the ground at the same time. • Show they have the same acceleration using F = ma. • Same acceleration - use equations of motion to show t is same provided heights are identical. • The smaller ball hits the ground first due to less air friction another possibility. • Difficulty releasing the balls at exactly the same time. • Dropped from the same height - bottom or centre of masses. • Dropped from the same height in a vacuum, air or fluid? • Balls of same material. • Comparison of surface areas.	3
7.	(a)	(i)		1
		(ii)	The forces acting on a unit positive charge (1) between the plates will be in the same direction (1) (no cancellation)	2
	(b)		The strong force (1) acting over a short range overcomes the electrostatic repulsion (1)	2
	(c)	(i)	2u 1d = (2 × 2/3) + (1 × − 1/3) = e (charge on proton)	1
		(ii)	1d 1 anti up quark	1
8.	(a)		200,000 J of energy transferred to each coulomb of charge	1
	(b)		Positive protons travel in the direction of the field or are attracted to negative tube/plate	1
	(c)	(i)	W = QV (1) $= 1{\cdot}6 \times 10^{-19} \times 200 \times 10_3$ (1) $= 3{\cdot}2 \times 10^{-14}$ J (1)	3
		(ii)	½ mv² = W (1) $½ \times 1{\cdot}673 \times 10^{-27} \times v^2 = 3{\cdot}2 \times 10^{-14}$ (1) $v = 6{\cdot}2 \times 10^6$ ms^{-1} (1)	3
	(d)		No effect, (1) since Q and V are constant (1)	2
9.	(a)		Constructive interference **OR** waves meet in phase **OR** crest meets crest and trough meets trough	1
	(b)	(i)	A Mean = (1·11 + 1·08 + 1·10 + 1·13 + 1·11 + 1·07) / 6 (1) = 1·10 m (1) B Random uncertainty = (1·13 − 1·07) / 6 (1) = 0·01 m (1)	4
		(ii)	$\% AB = \frac{0{\cdot}01}{1{\cdot}10} \times 100 = 0{\cdot}9\%$ (1) $\% BC = \frac{10}{270} \times 100 = 3{\cdot}7\%$ (1) BC has largest uncertainty (1)	3
		(iii)	$n\lambda = d\sin\theta$ (1) $2 \times \lambda = 4{\cdot}00 \times 10^{-6} \times 0{\cdot}270/1{\cdot}10$ (1) $\lambda = 4{\cdot}91 \times 10^{-7}$ m (1) $3{\cdot}7\%$ of $4{\cdot}91 \times 10^{-7} = 0{\cdot}18 \times 10^{-7}$ $l = (4{\cdot}91 \times 10^{-7} + 0{\cdot}18 \times 10^{-7})$ m (1) $(= (4{\cdot}9 \pm 0{\cdot}2) \times 10^{-7}$ m)	4
10.	(a)		$f = \frac{v}{\lambda}$ (1) $= \frac{3 \times 10^8}{605 \times 10^{-9}}$ $= 4.96 \times 10^{14}$ Hz (1) $E = hf_0$ $= 6{\cdot}63 \times 10^{-34} \times 4{\cdot}96 \times 10^{14}$ (1) $= 3{\cdot}29 \times 10^{-19}$ J (Answer given) **OR** $E = h\frac{v}{\lambda}$ (1) $= 6{\cdot}63 \times 10^{-34} \times 3 \times 10^8 / 605 \times 10^{-9}$ (1) $= 3{\cdot}29 \times 10^{-19}$ J (1)	3
	(b)	(i)	$E_k = (5{\cdot}12 \times 10{-}19) - (3{\cdot}29 \times 10^{-19})$ (1) $= 1{\cdot}83 \times 10^{-19}$ J (1)	2
		(ii)	Current reading decreases, (1) fewer photons hitting plate per second, fewer electrons released (1)	2
11.	(a)		$n = \frac{\sin\theta_a}{\sin\theta_g}$ (1) $1{\cdot}5 = \frac{\sin 50}{\sin\theta_g}$ (1) $\theta_g = 31°$ (1)	3

Question			Answer	Max mark
	(b)		$I = \frac{\lambda_a}{\lambda_g}$ (1) $1{\cdot}5 = \frac{\lambda_a}{420}$ (1) $\lambda_a = 630$ nm (1)	3
	(c)		Blue light has smaller wavelength than red in glass and so has a larger refractive index in glass **OR** Speed of blue light is slower than red in glass and so has a larger refractive index in glass	1
12.	(a)		$I = \frac{V}{R}$ (1) $= \frac{12}{480000}$ (1) $= 2{\cdot}5 \times 10^{-5}$ A (1)	3
	(b)		$V_C = 12 - 3{\cdot}8 = 8{\cdot}2$ V (1) $Q = CV$ (1) $= 2200\ \ 10^{-6} \times 8{\cdot}2$ (1) $= 1{\cdot}8 \times 10^{-2}$ C (1)	4
	(c)		$E = \frac{1}{2} CV^2$ (1) $= \frac{1}{2} \times 2200 \times 10^{-6} \times 12^2$ (1) $= 0{\cdot}16$ J (1)	3
13.	(a)		$V_p = 3 \times 0{\cdot}5 = 1{\cdot}5$ mV	1
	(b)		$f = \frac{1}{T}$ (1) $= \frac{1}{4 \times 10^{-3}}$ (1) $= 250$ Hz (1)	3
14.	(a)		**Open ended - see general marking instructions for breakdown.** Possible: • R_1 — charging • R_2 — discharging R_2 larger than R_1 • **Connected to A** — C charges through R_1, large current initially (less opposition) — current decreases as C fills up (increased opposition) • Electrons flow onto the plates • **Connected to B** - C discharges though R • Electron flow changes direction away from the plates — below the time axis on graph • Rate of flow is less due to higher value of R_2 • Charging rate faster than discharging rate	3

Question			Answer	Max mark
	(b)		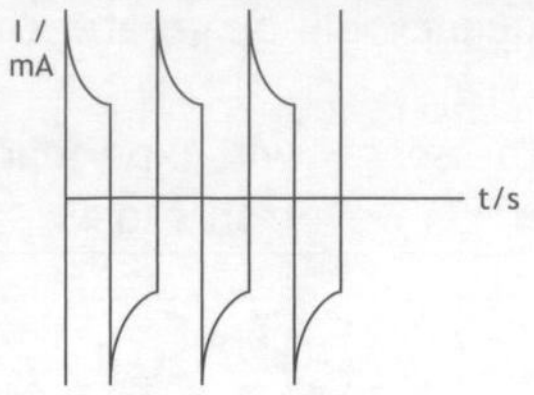 **shorter time interval** (1) **reduced fall in current** (1)	2

HIGHER FOR CfE PHYSICS MODEL PAPER 2

SECTION 1

Question	Answer	Mark
1.	A	1
2.	D	1
3.	B	1
4.	C	1
5.	E	1
6.	C	1
7.	C	1
8.	E	1
9.	B	1
10.	D	1
11.	B	1
12.	C	1
13.	B	1
14.	E	1
15.	A	1
16.	A	1
17.	A	1
18.	D	1
19.	D	1
20.	D	1

SECTION 2

Question			Answer	Max mark
1.	(a)		$F = \frac{Gm_1m_2}{r^2}$ $= \frac{6{\cdot}67 \times 10^{-11} \times 6{\cdot}0 \times 10^{24} \times 298}{(245 \times 10^3)^2}$ $= 2{\cdot}0 \times 10^6$ N	3
	(b)		$T = 2\pi\sqrt{\frac{r^3}{GM}}$ $24 \times 60 \times 60 =$ (1) $2\pi\sqrt{\frac{r^3}{6{\cdot}67 \times 10^{-11} \times 6{\cdot}0 \times 10^{24}}}$ (1) $r = 4{\cdot}24 \times 10^7$ m (1) Height above earth = $4{\cdot}24 \times 10^7 - 6{\cdot}4 \times 10^6$ m = $3{\cdot}6 \times 10^7$ m (1)	4
	(c)		Planets, constellations orbiting due to gravitational force. Centre of mass found to be much smaller than required (1) There would appear to be "missing mass" (1) – termed dark matter which cannot be seen	2
2.	(a)		A – hot, bright B – Cool, bright C - Cool, dim D - Hot, dim (2) for all correct (1) for two correct	2
	(b)		Light from stars analysed (1) Shorter wavelengths/higher frequencies have a greater temperature than longer wavelengths/lower frequencies (1) **OR** Violet end – hot and Red end – cooler (1)	2
3.	(a)		Component of weight = mgsinθ (1) = 60 × 9·8 × sin22 (1) = 220 N (1)	3
	(b)		Unbalanced force = 220 – 180 = 40 N (1) F = ma (1) 40 = 60 × a (1) a = 0.67 ms^{-2} Answer given	3
	(c)		$v^2 = u^2 + 2as$ (1) $= 0 + (2 \times 0{\cdot}67\ 50)$ (1) v = 8.2 ms^{-1} (1) **OR** $E_W = E_k$ $Fs = ½ mv^2$ (1) $40 \times 50 = ½ \times 60 \times v^2$ (1) v = 8·2 ms^{-1} (1)	3
	(d)		Smaller speed at bottom (1) Smaller component of weight down slope, smaller unbalanced force, smaller acceleration. (1)	2
4.	(a)	(i)	Total momentum before a collision equals the total momentum after the collision, **in the absence of external forces.**	1
		(ii)	Mom before = mom after *p = mv* (1) *(Can use $m_Au_A + m_Bu_B = (m_A + m_B)v$)* (0·22 × 0·25) + 0·16u = (0·38 × 0·2)(1) u = 0·13 ms^{-1} (1)	3
	(b)		Less, (1) total initial momentum is less, mass is constant. v = momentum/mass. (1)	2
5.	(a)	(i)	$v^2 = u^2 + 2as$ (1) $0 = 7^2 + 2 \times (-9{\cdot}8) \times s$ (1) s = 2·5 m (1)	3
		(ii)	v = u + at (1) 0 = 7 + (–9·8) t (1) t = 0·71 s Answer given	2
	(b)	(i)	At t = 0·71s, vertical velocity = 0 Horizontal velocity = constant = 1.5 ms^{-1} to the right.	1

Question			Answer	Max mark
		(ii)	Statement Z (1) The horizontal velocity of the ball is constant and equal to the velocity of the trolley. (1)	2
6.			**Open ended - see general marking instructions for breakdown.** Possible points: • Starts from rest • Possibility of labelling graph • Accelerates down slope • Loses energy during impact, v before greater then v after collision • Deceleration up slope is less than acceleration down slope • Unbalanced force greater on way down than up • Relate this to direction of the force of friction • Below axis indicates change in direction • Calculation of accelerations possible • Many possibilities available.	3
7.	(a)		<table><tr><td>Particle</td><td>Charge</td><td>Quark composition</td></tr><tr><td>Proton</td><td>+e</td><td>uud</td></tr><tr><td>Neutron</td><td></td><td>udd</td></tr><tr><td>Pion π^+</td><td></td><td>Up, anti down</td></tr></table>(1) for each correct answer	4
	(b)		Baryons consist of 3 quarks. (1) Mesons consist of 2 quarks. (1)	2
	(c)	(i)	They have the same mass but opposite charge. (1)	3
		(ii)	Positron, electron - any other suitable answers. (1)	
		(iii)	Annihilation or the release of energy. (1)	
8.	(a)		P = 40/20 = 2 m/W (1) $I = \frac{P}{A}$ (1) $= \frac{2 \times 10^{-3}}{8 \times 10^{-5}}$ (1) $= 25\ \text{Wm}^{-2}$ (1)	4
	(b)		Id^2 = 0·28, 0·32, 0·49 (1) = not constant so cannot be a point source. (1)	2

Question			Answer	Max mark
9.	(a)		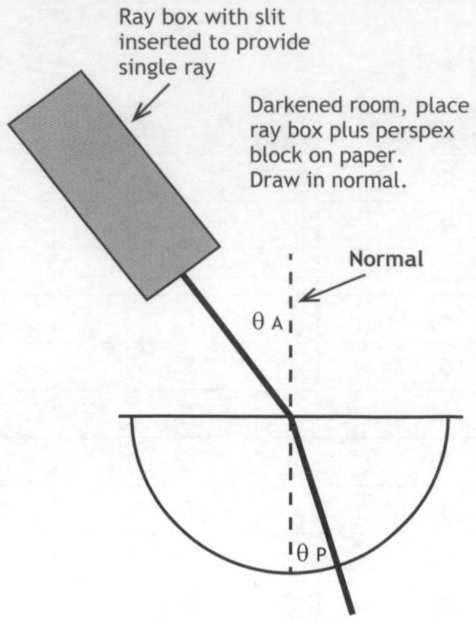 Ray box with slit inserted to provide single ray Darkened room, place ray box plus perspex block on paper. Draw in normal. Normal θ A θ P Correct angles plus normal (1) Ray box single slit (1) Block, A4 position (1)	3
	(b)		Measure the angle, θ_P, in perspex for corresponding angles θ_A of 10°, 20°, 30°, 40° and 50° or any other suitable five angles. (1) Repeat at least 3 times. (1) Calculate the mean values of θ_P and θ_A. (1)	3
	(c)		Note the scale reading uncertainty (± half the least scale division). (1) Calculate the random uncertainty for the three readings (Max – min) / 3 (1) (The largest value of uncertainty is the most significant).	2
	(d)		Calculate the values of sin θ_A and sin θ_P from the mean values (1) Plot a graph of sin θ_A against sin θ_P. (1) The gradient of the line gives the value of the refractive index. (1)	3
10.	(a)	(i)	Electrons cross the pn junction in the conduction band. (1) Once across they fall back to the valence band, releasing photons of light. (1)	2
		(ii)	$\lambda = \frac{v}{f}$ (1) $= \frac{3 \times 10^8}{6{\cdot}7 \times 10^{14}}$ (1) $= 4{\cdot}5 \times 10^{-7}$ m (1)	3

Question			Answer	Max mark
		(iii)	$E = hf$ (1) $= 6{\cdot}63 \times 10^{-34} \times 6.7 \times 10^{14}$ (1) $= 4{\cdot}44 \times 10^{-19}$ J (1) Caesium and strontium both emit photoelectrons. (1) **OR** Use $f_0 = \frac{E}{h}$ to calculate the threshold frequencies for each and compare with $6{\cdot}7 \times 10^{14}$ Hz Caesium $5{\cdot}1 \times 10^{14}$ Hz Strontium $6{\cdot}2 \times 10^{14}$ Hz Magnesium $8{\cdot}9 \times 10^{14}$ Hz (3) Caesium and strontium both emit photoelectrons. (1)	4
	(b)		$m\lambda = d\sin\theta$ (1) $2 \times 6{\cdot}35 \times 10^{-7} = 5{\cdot}0 \times 10^{-6} \times \sin\theta$ (1) $\theta = 14{\cdot}7^\circ$ (1)	3
11.	(a)		r = 95 (1) s = 7 (1)	2
	(b)		Total mass before is greater than total mass after reaction. (1) Mass loss converted to energy (using $E = mc^2$). (1)	2
	(c)		Total mass before = $3{\cdot}91848 \times 10^{-25}$ kg Total mass after = $3{\cdot}91478 \times 10^{-25}$ kg Change in mass = $3{\cdot}7 \times 10^{-28}$ kg (2) $E = mc^2$ (1) $= 3{\cdot}7 \times 10^{-28} \times (3 \times 10^8)^2$ (1) $= 3{\cdot}3 \times 10^{-11}$ J (1)	5
12.			**Open ended - see general marking instructions for breakdown.** Possible points: • Diffraction grating description / function • Interference pattern • Constructive interference - crest + crest / trough + trough • LED not monochromatic / small spread of wavelengths • Condition for maxima $d\sin\theta = m\lambda$ • Spread of pattern dependent on wavelength. • LED has an average wavelength greater than the laser. • LED — range of energy differences between the valence band to conduction band • Energy of emitted photons dependent on differences between valence and conduction bands. • Greater irradiance from laser light • Laser has a smaller cross sectional area.	3
13.	(a)	(i)	10 J of energy is given to each coulomb of charge by the supply.	1

Question			Answer	Max mark
		(ii)	r = Lost V/ I (1) = 2·5 / 1·25 (1) = 2 Ω. (Answer given)	2
	(b)	(i)	Total resistance has decreased, current has increased, (1) lost volts has increased, terminal potential difference has decreased. (1)	2
		(ii)	$E = IR_t + Ir$ (1) $10 = 2{\cdot}0\,R + 2{\cdot}0 \times 2$ (1) $R_t = 3\ \Omega$ (1) Resistance of parallel part of the circuit = 3Ω R = 6 Ω (Can use $\frac{1}{R_T} = \frac{1}{R_1} + \frac{1}{R_2}$) (1)	4
14.	(a)		V 0 t	1
	(b)		$V_R = IR$ (1) $= 5 \times 10^{-3} \times 500$ (1) = 2.5 V (1) $V_C = 12 - 2{\cdot}5$ = 9·5 V (1)	4
	(c)		$E = \frac{1}{2} CV^2$ (1) $= 0{\cdot}5 \times 47 \times 10^{-6} \times 12^2$ (1) $= 3{\cdot}4 \times 10^{-3}$ J (1)	3
	(d)		No effect (1) C and V remain unchanged. (1)	2

HIGHER FOR CfE PHYSICS MODEL PAPER 3

SECTION 1

Question	Answer	Mark
1.	E	1
2.	A	1
3.	D	1
4.	C	1
5.	D	1
6.	B	1
7.	A	1
8.	D	1
9.	E	1
10.	D	1
11.	D	1
12.	D	1
13.	A	1
14.	B	1
15.	D	1
16.	E	1
17.	B	1
18.	D	1
19.	A	1
20.	C	1

SECTION 2

Question			Answer	Max mark
1.	(a)		$\frac{\Delta\lambda}{\lambda} = \frac{v}{c}$ (1) $\frac{12}{586} = \frac{v}{3 \times 10^8}$ (1) $v = 6{\cdot}1 \times 10^6$ ms^{-1} (1)	3
	(b)	(i)	Calculate v/d for each value (15, 15·4, 15·4, 15) (1) so v/d = k (approx.) (1) so v α d (1)	3
		(ii)	Only 4 readings many more required **OR** no indication of uncertainty in each reading **OR** galaxies do not obey Hubble's law perfectly (other factors can effect rate of expansion).	1
	(c)		Age = 1 / Ho (1) = 1 / 2·3 x 10^{-18} s (1) = 1·4 x 10^{10} years (1)	3
2.	(a)		Component = mgsinθ (1) = 2600 × 9·8 × sin12° (1) = 5·3 × 10^3 N (1)	3
	(b)		Unbalanced force = ma (1) 5300 – 1400 = 2600a (1) a = 1.5 ms^{-2} (1)	3
	(c)		$v^2 = u^2 + 2as$ (1) = 5.0^2 + (2 × 1·5 × 75) (1) = 250 (ms^{-1})^2 $E_k = ½ mv^2$ (1) = ½ × 2600 x 250 (1) = 3·25 × 10^5 J (1)	5
3.	(a)	(i)	A Mean = $\frac{248 + 259 + 251 + 263 + 254}{5}$ = 255 μs (1) B Uncertainty = $\frac{263 - 248}{5}$ (1) = ± 3 μs (1)	3
		(ii)	Max time = 258 μs. (1) Club does not meet standard. (1)	2
	(b)	(i)	$F = \frac{mv - mu}{t}$ (1) $= \frac{4{\cdot}5 \times 10^{-2} \times (50 - 0)}{450 \times 10^{-6}}$ (1) = 5·0 × 10^3 N (1)	3
		(ii)	Greater impulse or greater change in momentum, (1) so greater speed (1)	2
4.	(a)		$v^2 = u^2 + 2as$ (1) $12^2 = 30^2$ + (2 × –9 × s) (1) s = 42 m (1) **OR** v = u + at 12 = 30 + 9t t = 2s s = ut + ½ at^2 (1) (30 × 2) + (1/2 × –9 × 2^2) (1) = 42 m (1) **OR** s = average v × t (1) = 21 × 2 (1) = 42 m (1)	3
	(b)		Speed at Q greater (1) mass of car greater, deceleration / acceleration less since a = F/m. (1) Could also do this by calculation.	2
	(c)	(i)	Electrons in the n type semiconductor combine with holes in the p type semiconductor. (1) Electrons fall from the conduction band to the valence band emitting energy as photons. (1) The energy drop dictates the frequency (colour) of the emitted light. (1)	3

Question			Answer	Max mark
		(ii)	$V_r = 12 - 5 = 7$ V (1) P = IV (1) 2·2 = I × 5 (1) I = 0·44 A V = IR (1) 7 = 0·44 x R R = 16 Ω (15·9 Ω) (1)	5
5.	(a)	(i)	s = vt (or area under graph) (1) = 20 × 3·06 = 61·2 m (1)	2
		(ii)	$v^2 = u^2 + 2as$ (1) 0 = 152 +(2 × −9·8 × s) (1) s = 11·5 m (1) **OR** s = area under graph (1) = ½ × 1·53 × 15 (1) = 11.5 m (1) **OR** s = ut + ½ at^2 (1) = 15 × 1·53 + ½ × = −9·8 × $(1{\cdot}53)^2$ (1) = 11·5 m (1)	3
	(b)		More likely to hit the tree (1) since horizontal velocity will decrease / horizontal range will decrease (1) **OR** time in air will decrease, (1) max height will decrease (1)	2
6.			**Open ended - see general marking instructions for breakdown.** Possible points: • Impulse Ft applied by goalkeeper. Force incorrectly used • Ball has to be stopped before going in opposite direction • Greater change in momentum • Time of contact greater compared to stationary ball • Force applied for a longer time - greater change in momentum. • Greater final speed, further distance • Angle of launch will also dictate distance travelled. • Ball compresses, elastic potential • Assuming contact at centre of ball • Possibility in decrease accuracy, spinning of ball	3
7.	(a)	(i)	$E_k = (hf - hf_o)$ $= 5{\cdot}23 \times 10^{-19} - 2{\cdot}56 \times 10^{-19}$ $= 2{\cdot}67 \times 10^{-19}$ J (1)	1
		(ii)	$E_k = ½ mv^2$ (1) $2{\cdot}67 \times 10^{-19} = ½ \times 9{\cdot}11 \times 10^{-31} \times v^2$ (1) $v = 7{\cdot}66 \times 10^5$ ms^{-1} (1)	3
	(b)		No change (1) — energy/frequency of the photons is constant (1)	2

Question			Answer	Max mark
8.	(a)		dsinθ = mλ (1) d × sin 35·3° = 3 × 633 × 10^{-9} (1) d = 3·29 × 10^{-6} m (1)	3
	(b)		No of lines per metre = 1 / 3·29 × 10^{-6} (1) = 3·04 × 10^5 m (1)	2
	(c)		Difference = (3·04 − 3·00) ×10^5 = 0·04 × 10^5 (1) Percentage difference = $\frac{0{\cdot}04 \times 10^5}{3{\cdot}00 \times 10^5} \times 100$ = 1·3% (1) Technician's value does agree. (1)	3
9.	(a)	(i)	E_3 to E_0 (1) Greatest energy difference has greatest f difference, (E = hf), but smallest wavelength, λ α 1/f. (1)	2
		(ii)	$W_3 - W_2$ (1) = (ΔE) = hf (1) $-5{\cdot}2 \times 10^{-19} - (-9{\cdot}0 \times 10^{-19}) = 6{\cdot}63 \times 10^{-34} \times f$ (1) f = 5.7 × 10^{14} Hz (1)	4
	(b)		$\lambda_A = \frac{v}{f}$ (1) $= \frac{3{\cdot}0 \times 10^8}{4{\cdot}6 \times 10^{14}}$ (1) = 6·5 × 10^{-7} m (1) $\frac{\lambda_a}{\lambda_g} = \frac{sin\theta_a}{sin\theta_g}$ (1) $\frac{6{\cdot}5 \times 10^{-7}}{\lambda_g} = \frac{sin\ 53°}{sin\ 30°}$ (1) $\lambda_g = 4{\cdot}1 \times 10^{-7}$ m (1)	6
10.	(a)		At A , $E_k = ½ mv^2$ (1) $= ½ \times 6{\cdot}64 \times 10^{-27} \times (2{\cdot}60 \times 10^6)^2$ $= 2{\cdot}24 \times 10^{-14}$ J (1) Increase in E_k = work done between plates $= 3{\cdot}05 \times 10^{-14} - 2{\cdot}24 \times 10^{-14}$ (1) $= 8{\cdot}1 \times 10^{-15}$ J (Answer given)	3
	(b)		W = QV (1) $8{\cdot}1 \times 10^{-15} = 3{\cdot}2 \times 10^{-19} \times V$ (1) V = 2·5 × 10^4 V (1)	3
	(c)		Since charge is less, less work is done, (1) so increase in E_k will be less (1)	2
11.	(a)		Upwards	1
	(b)		Gradient, m = 1·65 × 10^{-3} (Note F = × 10^{-3} N) (2) m = BI (1) from F = BIl 1·65 × 10^{-3} = B × 55 x 10^{-3} (1) B = 3·0 × 10^{-2} T (1)	5
	(c)	(i)	Scales not zeroed or wire not perpendicular to field or systematic uncertainty in the reading(s).	1
		(ii)	Repeat readings	1

Question			Answer	Max mark
12.			**Open ended – see general marking instructions for breakdown.** Possible points: • Solar wind – opposite charges • The earth's magnetic field will deflect these particles. • Deflected in different directions because of opposite charges • Some will get through to earth. • Follow field lines to poles. • Some follow circular paths / helical paths. • Borealis effects shown at poles when particles collide with our atmosphere. • Path of particles depend on angle depend on angle with magnetic field.	3
13.	(a)		The amount of charge that can be stored per volt **OR** the number of coulombs per volt.	1
	(b)	(i)	12 – 8.6 = 3.4 V	1
		(ii)	R = V / I = 3·4 / 0·0016 = 2125 Ω	3
		(iii)	V = 12 V (1) $E = \frac{1}{2}CV^2$ (1) $10{\cdot}8 \times 10^{-3} = \frac{1}{2} \times C \times 12^2$ (1) $C = 1{\cdot}5 \times 10^{-4}$ F (1)	4
	(c)		Time is less. (1) Circuit resistance is less or current / rate of flow of charge is larger (1)	2

HIGHER FOR CfE PHYSICS 2015

SECTION 1

Question	Answer	Mark
1.	C	1
2.	B	1
3.	A	1
4.	D	1
5.	C	1
6.	B	1
7.	C	1
8.	E	1
9.	D	1
10.	B	1
11.	A	1
12.	D	1
13.	D	1
14.	D	1
15.	A	1
16.	E	1
17.	B	1
18.	D	1
19.	E	1
20.	C	1

SECTION 2

Question			Answer	Max mark
1.	(a)	(i)	A $v = 11{\cdot}6$ ms^{-1} (1)	1
			B $v_h = 11{\cdot}6 \cos 40$ = 8·9 ms^{-1} (1)	1
			C $v_v = 11{\cdot}6 \sin 40$ = 7·5 ms^{-1} (1)	1
		(ii)	A $s = ut + \frac{1}{2}at^2$ (1) $4{\cdot}7 = 0 + \frac{1}{2} \times 9{\cdot}8 \times t^2$ (1) $t = 0{\cdot}979$ (s) (1) Total Time = 0·98 + 0·76 = 1·7 s (1)	4
			B $v = \frac{d}{t}$ (1) $8{\cdot}9 = \frac{d}{1{\cdot}7}$ (1) $d = 15$ m (1)	3
	(b)		kinetic energy is less (1) (as θ increases) speed decreases (1)	2

Question			Answer	Max mark
2.	(a)		(Total momentum before = total momentum after) $m_xu_x + m_yu_y = m_xv_x + m_yv_y$ (1) $(0{\cdot}25 \times 1{\cdot}20) + (0{\cdot}45 \times -0{\cdot}60) = (0{\cdot}25 \times -0{\cdot}80) + (0{\cdot}45 \times v_y)$ (1) $0{\cdot}30 - 0{\cdot}27 = -0{\cdot}20 + 0{\cdot}45 \times v_y$ $0{\cdot}45 \times v_y = 0{\cdot}23$ $v_y = 0{\cdot}51\ \text{ms}^{-1}$ (1) *(to the right)*	3
	(b)	(i)	*impulse = area under graph* $\left(= \frac{1}{2} b \times h\right)$ (1) $= \frac{1}{2} \times 0{\cdot}25 \times 4{\cdot}0$ (1) $= 0{\cdot}50$ N s (1) Accept 0·5, 0·500, 0·5000	3
		(ii)	0·50 kg ms^{-1} (1)	1
		(iii)	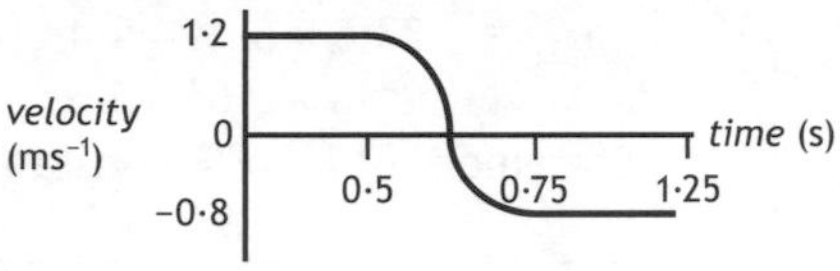 Constant velocity at correct values and signs before <u>and</u> after collision (1) Velocity change from initial to final in 0·25 s. (1) Shape of change of velocity correct ie initially gradual, increasing steepness then levelling out to constant velocity. (1)	3
3.	(a)		$F = \frac{GMm}{r^2}$ (1) $F = \frac{6{\cdot}67 \times 10^{-11} \times 6{\cdot}42 \times 10^{23} \times 5{\cdot}60 \times 10^{3}}{(3{\cdot}39 \times 10^{6} + 3{\cdot}70 \times 10^{6})^2}$ (1) $F = 4{\cdot}77 \times 10^3$ N (1)	3
	(b)		$g = \frac{W}{m}$ (1) $g = \frac{4770}{5600}$ (1) $g = 0{\cdot}852\ \text{N kg}^{-1}$ (1)	3
4.	(a)		photons of particular/some/certain energies/frequencies are absorbed (1) in its/the <u>Sun's</u> (upper/outer) atmosphere/outer layers (1)	2
	(b)	(i)	light is redshifted/ shifted <u>towards</u> red (1) (as) the galaxies are moving away (from the Sun) (1)	2
		(ii)	$z = \frac{\lambda_{observed} - \lambda_{rest}}{\lambda_{rest}}$ (1) $= \frac{450 \times 10^{-9} - 410 \times 10^{-9}}{410 \times 10^{-9}}$ (1) $= 0{\cdot}098$	2
		(iii)	$z = \frac{v}{c}$ (1) $0{\cdot}098 = \frac{v}{3{\cdot}00 \times 10^8}$ (1) $(v = 2{\cdot}94 \times 10^7\ \text{ms}^{-1})$ $v = H_0d$ (1) $2{\cdot}94 \times 10^7 = 2{\cdot}3 \times 10^{-18} \times d$ (1) $d = 1{\cdot}3 \times 10^{25}$ m (1) $(1{\cdot}4 \times 10^9$ ly)	5
5.			Demonstrates no understanding 0 marks Demonstrates limited understanding 1 marks Demonstrates reasonable understanding 2 marks Demonstrates good understanding 3 marks This is an open-ended question. **1 mark:** The student has demonstrated a limited understanding of the physics involved. The student has made some statement(s) which is/are relevant to the situation, showing that at least a little of the physics within the problem is understood. **2 marks:** The student has demonstrated a reasonable understanding of the physics involved. The student makes some statement(s) which is/are relevant to the situation, showing that the problem is understood.	3

Question			Answer	Max mark
			3 marks: The maximum available mark would be awarded to a student who has demonstrated a good understanding of the physics involved. The student shows a good comprehension of the physics of the situation and has provided a logically correct answer to the question posed. This type of response might include a statement of the principles involved, a relationship or an equation, and the application of these to respond to the problem. This does not mean the answer has to be what might be termed an "excellent" answer or a "complete" one.	
6.	(a)		Photon (1)	1
	(b)	(i)	126 GeV = $126 \times 10^9 \times (1{\cdot}6 \times 10^{-19})$ (1) $= 2{\cdot}0 \times 10^{-8}$ (J) $E = mc^2$ (1) $2{\cdot}0 \times 10^{-8} = m \times (3 \times 10^8)^2$ (1) $m = 2{\cdot}2 \times 10^{-25}$ (kg)	3
		(ii)	($2{\cdot}2 \times 10^{-25}/1{\cdot}673 \times 10^{-27}$ =) 130 (1) (Higgs boson is) <u>2</u> orders of magnitude <u>bigger</u> (1)	2
7.			Demonstrates no understanding 0 marks Demonstrates limited understanding 1 marks Demonstrates reasonable understanding 2 marks Demonstrates good understanding 3 marks This is an open-ended question. **1 mark:** The student has demonstrated a limited understanding of the physics involved. The student has made some statement(s) which is/are relevant to the situation, showing that at least a little of the physics within the problem is understood. **2 marks:** The student has demonstrated a reasonable understanding of the physics involved. The student makes some statement(s) which is/are relevant to the situation, showing that the problem is understood.	3

Question			Answer	Max mark
			3 marks: The maximum available mark would be awarded to a student who has demonstrated a good understanding of the physics involved. The student shows a good comprehension of the physics of the situation and has provided a logically correct answer to the question posed. This type of response might include a statement of the principles involved, a relationship or an equation, and the application of these to respond to the problem. This does not mean the answer has to be what might be termed an "excellent" answer or a "complete" one.	
8.	(a)		The power per unit area (incident on a surface)	1
	(b)		$134 \times 0{\cdot}2^2 = 5{\cdot}4$ $60{\cdot}5 \times 0{\cdot}3^2 = 5{\cdot}4$ $33{\cdot}6 \times 0{\cdot}4^2 = 5{\cdot}4$ $21{\cdot}8 \times 0{\cdot}5^2 = 5{\cdot}5$ (2) Statement of $I \times d^2$ = constant (1)	3
	(c)		$I \times d^2 = 5{\cdot}4$ (1) $I \times 0{\cdot}60^2 = 5{\cdot}4$ (1) $I = 15\ \text{W m}^{-2}$ (1)	3
	(d)		Smaller lamp (1) Will be more like a point source (1) Or Black cloth on bench (1) to reduce reflections (1)	2
	(e)		$A = 4\pi r^2 = 4\pi \times 2^2 = 50{\cdot}265$ (1) $I = \frac{P}{A}$ (1) $I = 24/50{\cdot}265$ (1) $I = 0{\cdot}48\ \text{W m}^{-2}$ (1)	4
9.	(a)	(i)	• Different frequencies/colours have different refractive indices (1) OR • Different frequencies/colours are <u>refracted</u> through different angles (1)	1
		(ii)	$n = \frac{v_1}{v_2}$ (1) $1{\cdot}54 = \frac{3.00 \times 10^8}{v_2}$ (1) $v_2 = 1{\cdot}95 \times 10^8\ \text{ms}^{-1}$ (1)	3

Question				Answer	Max mark
	(b)	(i)		$v = f\lambda$ (1) $3{\cdot}00 \times 10^8 = 4{\cdot}57 \times 10^{14} \times \lambda$ (1) $\lambda = 656{\cdot}5 \times 10^{-9}$ $m\lambda = d \sin\theta$ (1) $2 \times 656{\cdot}5 \times 10^{-9} = d \times \sin 19{\cdot}0$ (1) $d = 4{\cdot}03 \times 10^{-6}$ m (1)	5
		(ii)		• different colours have different λ (1) • $m\lambda = d \sin\theta$ (1) • (m and d are the same) • θ is different for different λ (1) OR • different colours have different λ (1) • Path difference = $m\lambda$ (1) • (for the same m) • PD is different for different λ (1)	3
10.	(a)	(i)		12·8 J (of energy) is gained by/supplied to 1 coulomb (of charge passing through the battery)	1
		(ii)		$E = V + Ir$ and $V = IR$ (1) $E = I(R + r)$ $12{\cdot}8 = I(0{\cdot}050 + 6{\cdot}0 \times 10^{-3})$ (1) $I = 230\ A$ (1)	3
		(iii)		(Wire of large diameter) has a low resistance (1) OR to prevent overheating (1) OR to prevent wires melting (1)	1
	(b)	(i)		12·6 V (1)	1
		(ii)		(gradient = $-r$) gradient = (12 − 12·5)/(60 − 10) (1) = −0·01 (1) internal resistance = 0·01 Ω (1)	3
		(iii)	(A)	$I = \frac{V}{R}$ (1) $= \frac{(15 - 11.5)}{(0.09 + 0.45)}$ (1) $= 6{\cdot}5\ A$ (1)	3
			(B)	The e.m.f. of the battery increases (1) Difference between the two e.m.f.s decreases (1)	2

Question			Answer	Max mark
11.	(a)		$C = \frac{Q}{V}$ (1) $64 \times 10^{-6} = \frac{Q}{2.50 + 10^3}$ (1) $Q = 0{\cdot}16(C)$	2
	(b)		$E = \frac{1}{2}QV$ (1) $E = \frac{1}{2} \times 0{\cdot}16 \times 2{\cdot}50 \times 10^3$ (1) $E = 200$ J (1)	3
	(c)	(i)	$V = IR$ (1) $2{\cdot}50 \times 10^3 = 35{\cdot}0 \times R$ (1) $R = 71{\cdot}4\ \Omega$ (1)	3
		(ii)	The voltage decreases (1)	1
		(iii)	Smaller initial current (1) Time to reach 0 A is longer (1)	2
12.	(a)		Suitable scales with labels on axes (quantity and units) (1) [Allow for axes starting at zero or broken axes or an appropriate value e.g. 30˚] Correct plotting of points (1) Smooth U shaped curve through these points. (1)	3
	(b)		36° and 66°	1
	(c)		37°	1
	(d)		Correct substitution into equation using D_m from answer to (c) (1) Correct value for n (1·5 if using D_m equal to 37°) (1)	2
	(e)		Repeat measurements (1) More measurements around/close to a minimum or smaller 'steps' in angle (1)	2

Acknowledgements

Permission has been sought from all relevant copyright holders and Hodder Gibson is grateful for the use of the following:

Image © Roman Chernikov/Shutterstock.com (SQP Section 2 page 12);
Image © Daseaford/Shutterstock.com (2015 Section 2 page 18);
Image © Dario Lo Pregti/Shutterstock.com (2015 Section 2 page 30).

Hodder Gibson would like to thank SQA for use of any past exam questions that may have been used in model papers, whether amended or in original form.